AF836031

Plant-Microbe Interactions in the Rhizosphere

Edited by

Adam Schikora
Julius Kühn-Institut, Braunschweig, Germany

Copyright © 2018
Caister Academic Press, U.K. www.caister.com

All rights reserved. No part of this publication may be reproduced, stored in a retrieval system, or transmitted, in any form or by any means, electronic, mechanical, photocopying, recording or otherwise, without the prior permission of the publisher. No claim to original government works. Ebooks are subject to the terms and conditions specified by the supplier.

ISBN: 978-1-912530-00-7 (paperback)
ISBN: 978-1-912530-01-4 (ebook)

Cover design adapted from images provided by Sven Jechalke and Adam Schikora.

Contents

Preface iv

1 You Are What You Can Find to Eat: Bacterial Metabolism in the Rhizosphere.......1
Nicola Holden

2 Role of Plasmids in Plant-Bacteria Interactions..17
Jasper Schierstaedt, Nina Bziuk, Nemanja Kuzmanović, Khald Blau, Kornelia Smalla and Sven Jechalke

3 Plant Immunity: The MTI-ETI Model and Beyond..39
Hanna Alhoraibi, Jean Bigeard, Naganand Rayapuram, Jean Colcombet and Heribert Hirt

4 Endofungal Bacteria Increase Fitness of their Host Fungi and Impact their Association with Crop Plants...59
Ibrahim Alabid, Stefanie P. Glaeser and Karl-Heinz Kogel

5 Plant-Nematode Interactions Assisted by Microbes in the Rhizosphere 75
Olivera Topalović and Holger Heuer

6 Apple Replant Disease: Causes and Mitigation Strategies 89
Traud Winkelmann, Kornelia Smalla, Wulf Amelung, Gerhard Baab, Gisela Grunewaldt-Stöcker, Xorla Kanfra, Rainer Meyhöfer, Stefanie Reim, Michaela Schmitz, Doris Vetterlein, Andreas Wrede, Sebastian Zühlke, Jürgen Grunewaldt, Stefan Weiß and Michael Schloter

Preface

Plants and the associated (micro)-organisms are in constant exchange of information. Those microorganisms may be associated with the host plant on different levels from facultative epiphytic or occasional colonisations to very intimate endophytic symbiosis. The notion that plants and associated microorganisms form stable communities promoted the use of the term holobiont to describe such multi-organismic systems. The exchanged information is usually encoded in different small molecules, polymeric structures and ions. Those are perceived by one or both partners and can influence very profoundly their physiology. The rhizosphere with its dense population of microorganisms is one of the so-called environmental hot-spots, in which interactions between the partners reach a very complex level.

The canonical definition of rhizosphere describes this part of the soil, which is directly influenced by the root system. This influence is the diffusion of root exudates such as sugars, acids, proteins and other low molecular weight organic compounds, together with the resulting chemical and physical changes of the soil structure. By definition the rhizosphere is limited on one side by the rhizodermis. However, since root cells are usually shed and replaced during growth and members of the endophytic community are regularly recruited from the soil, the actual border between the root and the rhizodermis is rather loose. Without doubt, the root surface and the adjoining soil layer are intensively colonized and both the quantity and quality of the colonizing microflora is controlled by the host plant to a significant extend.

This book gathers reviews and opinion papers on diverse aspects of the interactions which occur in the rhizosphere between the host plant and the microorganisms. The various reviews focus on particular phenomena, at the same time they represent the different levels of the interactions, from a biochemical and genetical basis to complex inter-organism communication.

Metabolism, "the underpinning force that sustains life", is the topic of the first chapter. The microbial community with its metabolic potential is shaped by the combination of plant genotype and the physiochemical properties of the soil. In addition, external influences such as climate, the degree of water saturation and anthropomorphic inputs influence metabolism throughout the rhizosphere community. The chapter describes the metabolic functions that occur in the rhizosphere either during bacteria-plant interactions or bacteria-bacteria interactions and discusses the mainly heterotrophic metabolism of organic substrates. The external influences mentioned above require a rapid adaptation to stresses and changing environmental conditions. Plasmids play an important role in bacterial adaptation. Their potential role is described in the second chapter. In the plant environment, plasmids can provide a selective advantage for the host bacteria, for example by carrying genes encoding metabolic pathways, metal and antibiotic resistances or pathogenicity related genes. This chapter provides an update on plasmids and horizontal gene transfer between plant-associated bacteria and their role in plant-bacteria interactions. Furthermore, it describes tools available to study the plant-associated mobilome. As well as microbial physiology and the flexible genetical equipment, the plant's response plays a crucial role in the interaction between the plant host and the associated microorganisms. One of the primary responses to microorganisms is a defense reaction termed microbe-associated molecular pattern (MAMP)-triggered immunity (MTI). Successful pathogens, however, can attenuate MTI by various effectors. This results in effector-triggered susceptibility and disease. Certain host plants have developed mechanisms to detect effectors and can trigger effector-triggered immunity (ETI), thereby diminishing pathogen propagation. Despite the wide acceptance of the above concepts, more and more accumulating evidence suggests that the distinction between MAMPs and effectors and MTI and ETI is often more complicated than originally thought. The following chapter discusses the complexity of MTI and ETI signaling networks and elaborates on the current definitions of MAMPs, effectors, MTI and ETI. It is clear that plant interactions with microorganisms are rarely (if ever) bilateral, rather interaction networks are the reality. One such example is described in the fourth chapter: a tripartite interaction between the host plant, its fungal endophyte and endofungal bacteria. There is increasing evidence that endofungal bacteria, alone or in combination with their fungal hosts, play a critical role in symbioses with plants. This chapter summarizes the current knowledge on endobacteria and their role in different types of fungal symbioses with plants. Since the frequency of bacteria in fungi is generally low, novel technology such as molecular taxonomy and advanced laser scanning microscopy were required to establish the functional contribution of these bacteria in those symbioses. In addition, isolation of those endobacteria permitted the study of the beneficial impact on plant hosts: plant growth promotion and resistance-inducing activity. Plant resistance to pathogens is indeed one of the central determinants of plant-microbe interactions and of the general health of the plant. The latter is strongly influenced by the interactions between pathogens and beneficial microorganisms. The fifth chapter focuses on soil suppressiveness, a phenomenon based on the impact of particular beneficial microorganisms on pathogens. Here, the authors describe soil suppressiveness as a biological tool against phytonematodes and explore the nature of monoculture versus crop rotation in this regard. Also, studies on the induction of the plant defence system and the establishment of so called induced systemic resistance by nematode-associated microbes (bacteria and fungi attached to phytonematodes) are discussed. Finally, this chapter discusses the importance of the knowledge on plant-nematode-microbe interactions in integrated pest management. The level of interaction relevant for agriculture is probably the most challenging to assess. To assure sufficient food production and high quality produce, farmers and relevant commercial companies need to manage quite complex situations. The sixth chapter describes a phenomenon, called apple replant disease (ARD). This is a poorly understood, soil-borne disease, resulting in severe growth suppression and decline in yield and fruit quality. The authors propose a new definition for ARD, highlighting its multiple causes including soil properties, faunal vectors, trophic cascades and genotype-specific plant secondary metabolism, particularly the biosynthesis of phytoalexins. Importantly, culture management with the emphasis on the improvement of soil microbial and faunal diversity as well as habitat quality rather than soil disinfection are suggested as a promising remedy. This accentuates even further the impact of interactions between (micro)-organisms and the plant on the overall plant performance.

Taken together, the exchange of information between the organisms in the rhizosphere not only influences the organisms themselves but also shapes the structure of the communities inhabiting this ecological niche. This impact is observable on different levels. The aim of this volume is to provide insights into phenomena which are exemplary for those diverse types of exchange. Since the advancement of technologies permits very fast progress in this field, it is very probable that the coming years will bring very detailed data on the interaction networks as well as on the possibility of managing the outcome of plant-microbe interactions for our benefit. This type of approach should be of great interest for new agricultural strategies and future plant protection practices.

Adam Schikora

CHAPTER 1

You Are What You Can Find to Eat: Bacterial Metabolism in the Rhizosphere

Nicola Holden*

The James Hutton Institute, Dundee, DD2 5DA, UK

*nicola.holden@hutton.ac.uk

DOI: https://doi.org/10.21775/9781912530007.01

Abstract
Metabolism is the underpinning force that sustains life. Within the rhizosphere it is a cyclic process, with substrates flowing between different compartments of the complete soil-plant-microbe system. The physiochemical and structural environment of the rhizosphere is shaped by a combination of plant genotype and soil type, both of which strongly impact the microbial community structure. External influences such as seasonality, the degree of water saturation and anthropomorphic inputs also play a role. Together these factors influence the flux of metabolites through the rhizosphere community, which in turn impacts on plant growth, development and disease. In this review, the focus is on metabolism within the bacterial population of the rhizosphere, since this group covers every type of plant-microbe interaction: from obligately symbiotic to destructively pathogenic, and includes those have little or no direct impact on plant hosts. The focus of the review is on metabolic functions that occur in the rhizosphere either during bacteria-plant interactions or bacteria-bacteria interactions and mainly covers heterotrophic metabolism of organic substrates. As such, many of the autotrophic (and phototrophic) reactions of inorganic compounds are not included.

Aim and Scope
The rhizosphere represents one of the most microbially diverse and dense habitats on the planet. The community structure of the rhizosphere varies widely, and largely is dependent on plant genotype and soil type, as well as being influenced by external climatic and anthropomorphic inputs. Microbiome analysis has highlighted wholescale differences in the rhizosphere microbiota from different ecosystems: for example contrast the communities derived from the pristine Amazonian rainforests with those from arid deserts or intensively cultivated agricultural regions. The aim of this review is to discuss how metabolic processes underpin the interactions of the bacterial population of the rhizosphere with their plant hosts, and examine the basis to the flux of metabolites in the rhizosphere system. Although there is not a clear delineation between the bulk soil and rhizosphere communities, this review is limited to metabolic functions that occur in the rhizosphere for bacteria that interact directly with plant hosts, or with other rhizospheric bacteria. As such, the review is focused on mainly organic substrates and many of the autotrophic and the phototrophic reactions for inorganic compounds are not included.

Metabolic pathways of bacteria
Broadly speaking, heterotrophic metabolism in bacteria is based on carbon- or nitrogen-containing organic compounds, i.e. carbohydrates (complexed or simple) and amino acids, whereas autotrophic metabolism is based on inorganic carbon or nitrogen. Bacteria can use multiple pathways to oxidise glucose: in addition to glycolysis, the hexose monophosphate shunt and the Entner-Doudoroff pathway can also be utilised, usually dependent on environmental parameters and lifestyle type. Some bacterial species lack key enzymes for all three pathways, such as *Azotobacter* and most *Pseudomonas* species that utilize the Entner-Doudoroff pathway for glucose catabolism, in contrast to the *Enterobacteriaceae*, as they lack the phosphofructokinase required for conversion to fructose (Jurtshuk P Jr., 1996) (T G Lessie and P V Phibbs, 1984). Metabolic flexibility is often associated with saprophytic pseudomonads due to the wide range of metabolites that they are likely to encounter in soil, exemplified by the bioremediation strain of *Pseudomonas putida*, which has been shown to have the capacity to use 911 metabolites (Nogales et al., 2008). Nitrate reduction in

heterotrophic bacteria is a feature of anaerobic metabolism that allows NO^{3}- ion to serve as a terminal electron acceptor (Jurtshuk P Jr., 1996), and the physiologic type of nitrate reductase forms the basis for metabolic separation of some species. While autotrophic bacteria still use CO_2 as a carbon source for growth, nitrogen is derived from NH_3, NO_3^- (nitrificiation and denitrification, respectively) or N_2 (nitrogen-fixing). A second group can oxidize sulphur compounds such H_2S, S_2, and S_2O_3, although they are not strictly autotrophic like some nitrifiers. Representatives of both groups are found in soil and can associate with the rhizosphere. However, a critical aspect of these reactions is that they are very sensitive to the presence oxygen and some such as denitrification and nitrogen-fixation, only occur under anaerobic conditions (Jurtshuk P Jr., 1996).

Functional plant-bacteria interactions

Rhizosphere-associated bacteria can be broadly split into functional groups dependent on how they interact with plant hosts, ranging from beneficial to overtly pathogenic. For those that have an intimately symbiotic relationship with the host, e.g. the rhizobia, the basis for the interaction is mutual exchange of substrates. This has been well described for fixation of atmospheric nitrogen and exchange of ammonia for carbohydrates. At the other end of the spectrum are the pathogens, although even for this group, the basis for development of disease is in the acquisition of nutrients from the plant host. Finally, in the middle are a diverse group classed as generalists or opportunists that perhaps receives less attention than the symbionts or pathogens. Cutting across these categorisations, different groups of rhizosphere bacteria are either shown to be nutritionally flexible or have evolved to exploit very specific ecological niches and associated substrates.

Beneficial interactors

Plant-microbe interactions that are classed as 'beneficial' normally relate to positive impacts on plant growth and development. This group of bacteria are generally termed plant growth promoting bacteria (PGP) and since the majority inhabit the rhizosphere or rhizoplane, they are termed PGP-rhizo-bacteria (PGPR). Metabolic flux frequently forms the basis to the interactions and whole system transcriptomic approaches have shown that over 50 % of the differentially expressed genes of respective bacteria and plant partners can be involved in metabolic processes, e.g. for a PGPR *Burkholderia* spp. and sugar cane host (Paungfoo-Lonhienne et al., 2016). One of the best-known beneficial interaction is in biological nitrogen fixation, where the basis of the symbiotic interaction is rooted in metabolism and exchange of nutrients. In essence, it involves encapsulation of rhizobia in bacteroids (specialised compartments in plant root cells), where they carry out the conversion of atmospheric nitrogen to ammonia, in exchange for plant-derived carbohydrates. Since it has been expertly explained in more detail elsewhere, the reader is referred to excellent reviews on nitrogen fixation such as (Dixon and Kahn, 2004; Hayat et al., 2010; Frans J. de Bruijn, 2015).

Others rhizospheric bacteria (herein termed rhizobacteria), such as PGPR do not have such an intimate dependency on their hosts, but can still play a beneficial role albeit with more complex functional interactions. A major taxonomic group in the rhizosphere are the Firmicutes, which includes the model Gram positive species, *Bacillus subtilis*. Members of the *Bacillus* genus are capable of functions that benefit the plant host directly, such as mineral nutrient solubilisation, e.g. through the production of enzymes that solubilise organic phosphorus, mainly stored as insoluble myo-inositol hexaphosphate or phytate, into a plant-usable form. They can also secret phytohormone mimics to induce plant growth and have indirect effects such as biocontrol and antibiosis that act on other members of the rhizosphere community in an antagonistic manner (Francis et al., 2010). Furthermore, PGPR can alter the plant defence response, which in turn impacts other members of the microbial community, again in a negative, competitive manner. Within this group, the spectrum of interactions is likely to be broad and specific to a particular bacteria-plant system, with different degrees of benefit to the host plant.

The pseudomonads comprise another major taxonomic group within the rhizosphere. Individual genotypes, even within a single species, are capable of a wide range of functional interactions with plant hosts, exemplified by *Pseudomonas fluorescens*. This is a diverse species that consisting of a complex of sub-species (Scales et al., 2014). On one hand, there are isolates that are known to promote plant growth (Silby et al., 2009; Redondo-Nieto et al., 2013) and some are available commercially as agricultural inputs. The basis to their functional activity, like some of the *Bacillus* species, is wide-ranging, dependent on a variety of traits including phytohormone production, biofilm formation, siderophore activity and direct anti-microbial competition via production of antimicrobial compounds. Isolate-specific differences are evident, e.g. in their ability to tap into phytohormone signalling, where some isolates encode genes for IAA synthesis, while others produce ACC deaminases that promote root elongation (Shen et al., 2013). At the other end of the spectrum, some *Ps. fluorescens* isolates are able to cause soft-rot disease via the production of secreted enzymes to degrade pectin and other cell wall components (Silby et al., 2011). Much attention has been given to this organism as a model species for different metabolic attributes, from nitrogen fixation (Haahtela et al., 1983) to denitrification (Redondo-Nieto et al., 2013), and its ability to generate similar levels of ATP under different substrate limitations and stresses (Silby et al., 2009; Appanna et al., 2016).

Opportunists
One group of bacteria appear to be related to the PGPRs because of their ability to fix atmospheric nitrogen, although the metabolic flux appears to be tipped in favour of the bacteria. They are distinct from the 'ineffective' rhizobia that either cannot or do not fix nitrogen in a mutualistic fashion and still infect legumes (Denison and Kiers, 2004), but instead exist as free-living dizatrophs. While there are those that clearly play a beneficial role, e.g. *Herbaspirillum* spp. for which fixation of nitrogen in association with non-legumous crop hosts has been demonstrated (James, 2000), there are others that for which the interaction is less likely to be of benefit to the plant hosts, e.g. for *Azotobacter* and *Klebsiella* spp.. *Klebsiella pneumonia* is probably best known for its ability to cause disease in humans as an opportunistic nosocomial pathogen, but some isolates are equally at home on plants (Brown and Seidler, 1973; Haahtela et al., 1986; Falomir et al., 2013) and other species, such as *K. oxytoca* are well known members of the soil and plant community (Bagley, 1985). Some isolates of *K. pneumonia* (and *K. oxytoca*) that have been isolated from plants encode genes for nitrogen fixation and are classed as diazotrophic, e.g. *K. pneumoniae* isolate Kp342 (Dong et al., 2003a; Fouts et al., 2008). Indeed, its interaction with plant hosts induces a host response (Iniguez et al., 2005) but the extent to which nutrient exchange takes place is less well established beyond an interaction with a particular variety of wheat (Red Baron) (Iniguez et al., 2004). It is possible that instead, these genes provide an extension of the bacterial metabolic capacity in environments where ammonia is lacking and insufficient oxygen induces an anaerobic physiological response, and thus play little or no beneficial role in the plant-bacteria interaction. Indeed, the minimal gene set for nitrogen fixation is present in at least 149 diazotrophic species (Dos Santos et al., 2012), including some species that have a well-characterised phytopathogenic interaction with the plant hosts, for example *Pectobacterium atrosepticum*, a species that causes devastating disease on potato (Bell et al., 2004). Together this supports the concept of an extended metabolic capacity that is induced under challenging physiochemical conditions.

K. penumoniae and other rhizosphere bacteria that are nutritionally flexible form a functional group that are defined as generalists or could be termed 'opportunistic'. Broadly speaking, they neither benefit nor harm the plant hosts, but instead are attracted to the rhizosphere as a nutrient-rich ecological niche. Bacteria within this group are made up diverse taxonomic groups and are capable of a broad range of metabolic functions, able to exploit not only plant-derived rhizodeposits, but substrates released by other members of the community (Badri et al., 2009; Bakker et al., 2013). Other members of this group include some species that although they are likely to be present at very low densities, garner a good deal of attention. These include bacteria that cause disease to human and animal hosts, e.g. pathogenic *Escherichia coli* and non-typhoidal *Salmonella enterica* (Holden et al., 2009; Holden et al., 2015).

Plants were not considered to be hosts for bacteria that are commonly associated with animals until relatively recently. It was originally thought that foodborne illness as a result of consumption of contaminated fruit and vegetables had only arisen through transient transmission of foodborne pathogens on the food products. This would be akin to viruses and parasites that can only complete their replication cycle within a susceptible animal host, but have still been shown to be transmitted into the food chain via edible crops (Chancellor et al., 2006; Amoros et al., 2010; Macarisin et al., 2010; Made et al., 2013; Swinkels et al., 2014; Einoder-Moreno et al., 2016). However, numerous studies have shown that human pathogenic bacteria can interact with plants and use them as hosts, which by definition requires cell mass growth and division (Holden et al., 2015). Given the richness of species associated with the rhizosphere, it is perhaps not surprising that it was recognised as a reservoir for human pathogenic bacteria (Berg et al., 2005), although there are differences in colonisation ability between the species, e.g. for a lab-adapted isolate of *E. coli* that compares relatively poorly to *S. enterica* serovar Cubana or an endophytic isolate of *K. pneumoniae* (Dong et al., 2003b). Colonisation in the opposite direction, from plants to humans has also been reported, e.g. for some isolates of well-characterised rhizobacteria such as *Ps. fluorescens* that behave as opportunistic human pathogens (Scales et al., 2014), and other human pathogens such as *Listeria monocytogenes* and *Pseudomonas aeruginosa* that are considered to be 'environmental' or soil-derived (Stover et al., 2000; Freitag et al., 2009). Therefore, these bacteria also fit within the group of rhizosphere metabolic opportunists.

E. coli and *S. enterica* are well-characterised model bacteria (Neidhardt and Curtiss, 1999) for which their metabolic capabilities have been thoroughly researched and are probably the best known of all bacteria. Given their metabolic flexibility, these bacteria can also be classed as 'opportunistic' colonisers of plants. However, much of what we have learnt has been derived at temperatures that are relevant to mammalian hosts, which can be 20 °C higher or more than edible crops grown in temperate zones, thereby raising questions about relevance. Nonetheless, these are mesophilic bacteria capable of growth over a wide temperature range encompassing most plant growth requirements. Modelling based on multiple genomes of *E. coli* isolates, including established pathotypes has resulted in the generation of metabolic models for a pangenome and a core genome (Baumler et al., 2011; Monk et al., 2013). The *E. coli* EHEC pathotype that belongs to foodborne group and are frequently associated with contamination of edible crop species (Holden et al., 2009), encode unique genes for metabolic reactions that are relevant to plant-derived substrates, including salicylate hydroxylase, gentisate 1,2,-dioxygenase, sucrose transport and fucose synthetase (Baumler et al., 2011). The function of these gene products points to a role for catabolism of plant derived metabolites and although *E. coli* are normally associated with vegetarian or omnivorous mammals and hence have access to plant-derived material in the animal gut, there is an obvious need to understand their response to plant hosts directly.

Phytopathogens

Research on plant pathogens generally focuses on the mechanisms of disease and how it can be prevented or reduced. As with pathogens of animals, it accepted that the underlying driver of pathogenesis is metabolism as a means to exploit the host as a source of metabolic substrates. This can be manifested by phytopathogens in the production of specialised enzymes to break cell walls, or in manipulation of the host defence to facilitate colonisation. The vast majority of characterised phytopathogens cause symptomatic damage to aerial plant tissue, e.g. fruit bodies or leaves and as such any reported interaction within the rhizosphere is either absent or if it does occur, is overlooked. One group that is recognised to cause damage to below-ground tissues are the soft-rot erwiniae, which cause a major economic threat to potato stolon production. The soft-rot erwiniae produce a plethora of enzymes, targeting not just pectin, but also cellulose in the plant cell wall (Harris et al., 1998). In fact, this metabolic function has been exploited in biotechnology as part of co-cultures used to reduce recalcitrant substrates to bio-ethanol (Grohmann et al., 1998). Development of symptomatic disease from cell-wall degradation has been shown to be a direct response to population density, e.g. for *Pectobacterium atrosepticum*, and production of cell-wall degrading enzymes only occurs in a quorum-dependent manner when the population becomes sufficiently large that there is a need to increase acquisition of nutrients through the active release of cytoplasmic contents (Toth and Birch, 2005). In a similar manner, quorum sensing-dependent colonisation and pathogenicity has been shown to occur for *Ps. aeruginosa*, resulting in necrotic lesions on the root tips of sweet basil and *Arabidposis thaliana* (Walker et al., 2004).

Plants are not without their own defences, and responses to bacterial molecular patterns has been well established as a first-line defensive strategy, followed by more specific responses that counteract the function of bacterial effectors (Jones and Dangl, 2006). As part of the response, plants also produce secondary metabolites as active anti-microbial compounds, such as the inhibition of quorum sensing, which could prevent phytopathogens reaching numbers that cause cellular destruction. One of these is a phenolic antimicrobial compound, curcumin, which at sub-inhibitory concentrations has been shown to reduce *Ps. aeruginosa* (isolate PAO1) pathogenicity on *A. thaliana* by interfering with quorum sensing (Rudrappa and Bais, 2008). However, it is notable that a pathway for curcumin catabolism via a curcumin/dihydrocurcumin reductase has been identified in *E. coli* (Hassaninasab et al., 2011), and indeed this enzyme was found to be induced in *E. coli* O157:H7 in response to spinach root exudates (Crozier et al.,

2016). The genes for degradation of gentisate, a product of salicylic acid, were similarly induced. This raises the intriguing possibility that opportunists such as *E. coli* are able to exploit aspects of plant immunity for metabolism, although this has yet to be shown.

Metabolism of plant-associated bacteria is an active area of research and has reached the stage where metabolic networks can be constructed. These range from research for biotechnology to a fundamental understanding of the pathways involved. Examples include the identification of bactericides from *Pectobacterium carotovorum* for agricultural use (Wang et al., 2015); identification of differences between pathogenic and non-pathogenic pseudomonads (Mithani et al., 2011); the process of denitrification using *Ps. fluorescens* as a model (Arat et al., 2015); and for bioethanol production from plant-based feedstocks for *B. subtilis* (Qi et al., 2014). Metabolic pathway analysis can be reconstructed from gene expression data to examine substrate utilisation for the foodborne pathogen *E. coli* O157:H7 with different edible crop tissue types, such as spinach root exudates (Figure 1).

Role of the plant host

Technical advances in sequencing technology have facilitated detailed community and microbiome analysis, which is revolutionising microbiology. With the appropriate approaches and analysis, a more detailed picture of plant-microbe interactions is emerging, at not just a taxomonic but also functional level. What has become clear is that the rhizosphere-associated microbiota effectively increase the functional diversity of the rhizosphere by orders of magnitude (Bakker et al., 2013). One of

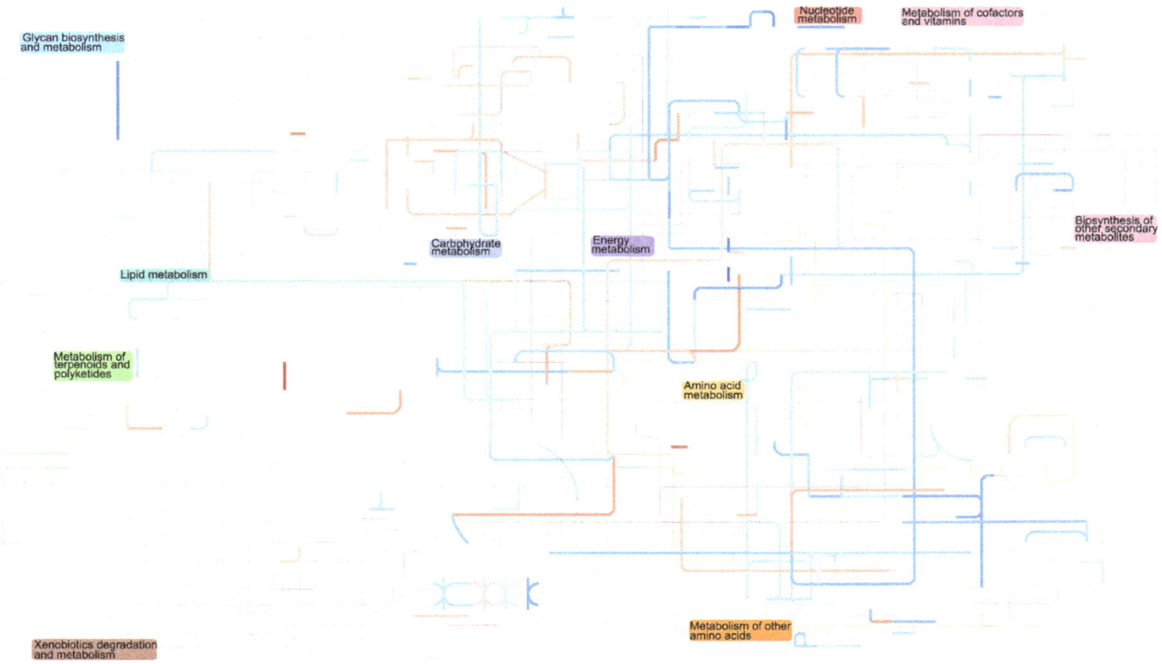

Figure 1. KEGG pathway analysis of *E. coli* O157:H7 transcriptomic response to spinach root exudates.
Gene expression microarray data derived from incubation of *E. coli* O157:H7 (isolate Sakai) in spinach root exudates for 1 hour (Crozier et al., 2016) was analysed with the KEGG pathways tools (Okuda et al., 2008). Genes that correspond to KEGG orthologs were mapped to the metabolic pathways KEGG map for *E. coli* K-12 (reference isolate MG1655). The fold-change gene expression data (i.e. exudates compared to the no-plant control) were log-transformed and scaled to a red (high) / blue (low) colour scheme. All other reactions/compounds were greyed-out (unpublished, courtesy of Leighton Pritchard, James Hutton Institute).

these key aspects is driven by bacterial metabolism (Hacquard et al., 2015). One of the most profound findings is the strength of influence that rhizodeposits exert on microbial community composition and hence function (Wagner et al., 2016). Substrate-driven community recruitment has been reported for model species, *A. thaliana,* as well as non-model species such as wild oat (*Avena fatua*) (Bulgarelli et al., 2013). Microbiome analysis of the rhizosphere from different accessions of barley (wild and domesticated), showed a significant host genotype-dependent impact. However, the difference between barely genotypes was not particularly large, presumably because of their recent ancestry, in comparison to larger Family-level differences comparing barley with *A. thaliana* (Bulgarelli et al., 2015). Analysis of the functional groups in the barley rhizosphere community revealed enrichment in genes involved in plant-microbe interactions, e.g. in virulence and adherence, but also in siderophores and phosphotransferase systems (Bulgarelli et al., 2015). Logical questions that arise from such community-level research are in availability and localisation of plant-derived substrates.

Substrate availability

Polysaccharides

Various techniques have been applied to taking a bottom-up (genes to population) approach to identify the genetic basis to colonisation of the rhizosphere. One of the earlier reports took a positive selection approach, using *in vivo* expression technology, to show the contribution of metabolism-associated genes for *Ps. fluorescens* during colonisation of rhizosphere of sugar beet seedlings (Rainey, 1999). In a similar manner, whole genome transcriptomic analysis has been carried out to determine the response of foodborne pathogens to plants or plant extracts, focusing almost entirely on *E. coli* O157:H7 (Bergholz et al., 2009; Kyle et al., 2010; Fink et al., 2012; Hou et al., 2012; Hou et al., 2013; Landstorfer et al., 2014; Crozier et al., 2016; Linden et al., 2016). Although this global gene expression approach enables reconstruction of metabolic pathways, it is surprisingly difficult to find consistencies between the responses. This is likely due to differences in the experimental set-up that alter the physiochemical environment, one of the most important aspects of which is temperature. However, together the data support the view that these bacteria can function as opportunistic plant colonisers and that there is specificity in their response, dependent on the plant species and tissue type.

The plant cell wall represents a major evolutionary difference between plants and animals and as a consequence, genomic differences that relate to utilisation of cell-wall metabolites can serve as a basis for differentiation of plant- and animal-association of bacteria. The differences between the animal and plant kingdoms that relate to cell wall glycans as potential metabolites are in their mechanism of secretion; the linkages between glycans; and the glycans themselves. In plants, unique modification of *N*-glycans in the Golgi occur at the trimming stage (Etzler ME, 2009). After the addition of N-acetylglucosamine to the distal mannose of the core by *N*-acetylglucosamine transferase I (GnT-I), two specific plant modifications occur: the addition of xylose in β1-2 linkage to the core β-mannose, which is unique to plants; and the addition of fucose in α1-3-linkage to the asparagine-linked N-acetylglucosamine residue, which has also been found in invertebrates (Etzler ME, 2009). L-arabinose, like xylose, is a pentose that is unique to plants and green algae, and depending on the plant species, can account for 5 - 10 % of plant cell wall glycans (Kotake et al., 2016). It is mainly located in pectin as side chains on rhamnogalacturonan I (RG-I) and rhamnogalacturonan II (RG-II), and in arabinogalactans, *O*-linked hydroxyproline-rich glycoproteins (Mohnen, 2008). In addition to the presence of xylose on complex or hybrid *N*-linked glycoproteins, it is also present as xyloglucan, the major component of hemicellulose in higher plants, and as a minor component of pectin in xylogalacturonan (Etzler ME, 2009). Biosynthesis of L-arabinose occurs from epimerization of UDP-D-xylose (Burget et al., 2003). The action of phytopathogen cell-wall degrading enzymes provides access to this resource, e.g. the xylanolytic system, which is ubiquitous in lignocellulose-degrading microbes and has been well-characterised for the phytopathogenic xanthomonads (Santos et al., 2014).

In bacteria, metabolism of xylose is intrinsically linked to arabinose, since the low affinity transporter (AraE) can accommodate either pentose, and the master regulator (AraC) controls expression of both sets of metabolic genes, as shown for *E. coli* (Desai and Rao, 2010). Catabolite repression results in hierarchical utilisation of firstly arabinose followed by xylose. In *E. coli* O157:H7 the *araBAD* and *xylBA* loci were upregulated in response to spinach root exudates after just one hour (Crozier et al., 2016), but downregulated after two days colonisation of lettuce roots (Linden et al., 2016), indicating a transient and dynamic response. Functional ability to catabolise arabinose has also been demonstrated during colonisation of the pea root rhizosphere by *Rhizobium leguminosarum* biovar *viciae*, using a signature-tagged mutagenesis approach that identified essential rhizosphere-colonisation genes (Garcia-Fraile et al., 2015). *Ps. fluorescens* colonisation of rhizosphere of sugar beet seedlings resulted in induction of genes for metabolism of xylose as well as complex N-compounds and sensory systems (Rainey, 1999), although subsequent characterisation of the genes revealed specificity in their functional roles on different species, e.g. for histidine sensing and utilisation (Zhang et al., 2006).

Amino acids and cyclic compounds

Roots secrete an array of biologically active compounds, including ions, water, oxygen, enzymes, mucilage and primary and secondary metabolites (Bais et al., 2006). This provides an enviable niche inhabited a rich diversity and density of microoganisms (Bakker et al., 2013). The rhizosphere of numerous plant species has been shown to support the growth of foodborne pathogenic bacteria, (Jablasone et al., 2005; Ibekwe et al., 2009; Quilliam et al., 2012; Wright et al., 2013) (Wright *et al* MBT paper in press)(Semenov et al., 2010; Kisluk and Yaron, 2012; Mendes et al., 2013). Root exudates play an important role in attracting bacteria to the rhizosphere and are often metabolites in their own right (Huang et al., 2014). Amino acids released from alfalfa seedling exudates (tryptophan, methionine, lysine, and phenylalanine) were found to be depleted in the presence of *S. enterica* indicative of *de novo* amino acid metabolism (Kwan et al., 2015). Amino acid transport and metabolism has been shown to be one of the largest classes of shared genes in catabolism of plant-derived substrates, in a genomic comparison of PGPR *Ps. fluorescens* isolates. For three isolates in the *Ps. fluorescens* species complex (i.e. not *sensu scritro*) (GP72, Pf-5 and M18), the number of shared genes exceeds 500 (Shen et al., 2013). Genes for the catabolism of aromatic amino acids phenylalanine and tyrosine, as well as oxygenases for other aromatic compounds are well represented in the *Ps. fluorescens* group, and to a lesser extent also in the related rhizobacteria, *Ps. pudita* (Shen et al., 2013).

Micronutrients

PGPR bacteria can contribute to plant growth through the solubilisation of phosphorus into plant-available forms. For *Ps. fluorescens* isolates, this has been demonstrated from enzyme activity of non-specific phosphatases, inositol phosphate phosphatases and C-P lyases (Shen et al., 2013). Similar functional solubilisation has been shown for Gram-positive PGPR, including isolates from *Bacillus*, *Brevibacterium*, *Sarcina*, *Paenibacillus*, *Corynebacterium* and *Micrococcus* species (Francis et al., 2010).

Iron limitation is known to be a factor affecting colonisation of plant-associated bacteria, e.g. *Ps. aeruginosa* (Sulochana et al., 2014), *Erwinia amylovora* (Smits and Duffy, 2011) and *Ps. syringae* pv. *tabaci* (Taguchi et al., 2010). Iron was found to be depleted in alfalfa exudates in the presence of *S. enterica* and correlated with increased enterobactin expression (Hao et al., 2012). A similar limitation of Fe^{3+} was apparent for *E. coli* O157:H7 inoculated in root spinach exudates, resulting in induction of systems for ferric iron and haem transport, via enterobactin and Chu transport system respectively (Crozier et al., 2016). These studies have relied on extracts or plants grown under hydroponics systems, and there are inherent technical difficulties for such molecular mechanistic data from the roots and rhizosphere of soil/compost-grown plants. However, competition for iron has been postulated to be one of the possible mechanisms for the beneficial effects demonstrated by PGPR on phytopathognic microbes (Wensing et al., 2010; Yu et al., 2011).

Excess micronutrients, such as heavy metals, can occur from anthropomorphic activities resulting in pollution. Plants that are capable of growing in polluted sites are termed metal hyperaccumulators, and are dependent on the metabolic functions of rhizobacteria for their resistance (Visioli et al., 2015). There are various examples of heavy metal accumulation with some evidence for links to the plant defence response by incorporation of the compounds to levels that are toxic to phytopathogens (Fones and Preston, 2013). There are also strong drivers within industrial biotechnology to identify rhizobacteria that can aid in bioremediation. For example, mobilisation of copper by a PGPR species, *Ps. putida* has been demonstrated in *Elsholtzia splendens*, a hyperaccumulator used in phytoremediation, by promoting copper redistribution in the plant root and translocation from root to shoot (Xu et al., 2015).

Substrate sensing

Metabolism and growth potential are inherently tied in to responsive changes in gene expression. Despite the nutritional abundances in the rhizosphere, it is a limiting environment in some respects, inducing physiological stress responses and resulting in restricted growth rates. Niche separation or partitioning is an accepted concept in soil microbiology (Lennon et al., 2012), and growth characteristics of soil bacteria have shown distinct phyla and order level differences in their utilisation of labile and recalcitrant carbon-based substrates (e.g. glycine / sucrose 'vs' cellulose / lignin / tannin-protein) (Goldfarb et al., 2011). As such, successful colonisation and establishment requires responsive sensing of the surrounding environment to obtain physical access to the appropriate substrates, followed by effective and efficient re-routing of the metabolic circuitry to best exploit the resources.

Chemotaxis

Bacteria respond to the presence of a substrate by moving down a gradient toward it. For the majority of species, this motility is flagella-driven via induction of chemotaxis genes (Sourjik and Wingreen, 2012), and a role for flagella in plant-associated bacteria has been well established in promoting root colonisation, e.g. for *Ps. fluorescens* (De Weger et al., 1987). In symbiotic bacteria, there are a number of chemoreceptors that aid targeting of the bacteria host interaction, for example with *Sinorhizobium meliloti* in sensing its host *Medicago sativa* (Webb et al., 2016). Once the interaction has been established and bacteria growth commences, this function is down-regulated, as shown for *Ps. fluorescens* growing in the presence of sugarbeet (*B. vulgaris* L.) root exudates (Mark et al., 2005). The presence of flagella was shown to contribute to survival of *S. enterica* serovar Dublin in manure-amended soil (Olsen et al., 2012) and a chemotactic response occurred for *S. enterica* serovar Typhimurium on lettuce leaves (Kroupitski et al., 2009). Both serovars, Typhimurium and Dublin were shown to exhibit chemotaxis towards, and induction of metabolism genes in response to the presence of root exudates from lettuce plants (Klerks et al., 2007).

The concept of direct sensing or responding to physiochemical changes has been well documented with description of riboswitches. These are natural aptamers imbedded in the gene leader sequences (5'UTR) that selectively bind small molecules such as second messengers, and have been reported for numerous metabolic genes, e.g. vitamins and amino acids (Nudler and Mironov, 2004). In a recent study in *Ps. putida*, *in vivo* characterisation of the riboswitch that binds the active form of thiamine (vitamin B1, TPP) to represses the expression of thiamine-related genes, showed that the riboswitch acted at the translational level by interfering with RBS-ribosome recognition (D'Arrigo et al., 2016). Riboswitch function has been exploited in synthetic biology for a range of uses, including the potential removal of agri-chemicals or other pollutants. The chemotaxis locus *cheX* in *E. coli* was reprogrammed to respond to the herbicide atrazine, and incorporation of genes for metabolism of atrazine generated a strain that is capable of sensing, chemotaxis and degradation of the compound (Sinha et al., 2010).

Two-component systems

Two-component systems are the primary mechanism for sensing substrates. They act at the transcriptional level, via sensing and signal transduction, often involving a cascade of regulation from the global transcriptional response-regulator. For example, the *barA-uvrY* system (and its orthologues) is widespread in bacteria and is involved in the regulation of multiple responses including metabolism (Sahu et al., 2003). In the γ-Proteobacteria it controls expression of the carbon storage regulator (*csr*) / repressor of stationary phase metabolites (*rsm*) system, a widespread regulatory metabolic network, by binding regulatory elements in an antagonistic manner (Zere et al., 2015). A role for *rsm* has been shown for *Ps. fluorescens* (Valverde et al., 2004), but it is generally reported for this genus in the context of biofilm formation, for example for a plant-growth promoting isolate of *Ps. putida* (Huertas-Rosales et al., 2016). Another well described two-component regulatory system is *ntrB-ntrC*, which is activated under nitrogen limiting conditions for nitrogen uptake and metabolism. In *Ps. fluorescens*, its link to motility was elegantly demonstrated in an experimental evolution approach using a non-motile mutant that lacks both flagella and viscosin motility. Following repeated culturing on motility medium the strain became partly motile via mutation in the *ntrB* gene, which resulted in over-activation of nitrogen regulation, uptake and metabolism genes. This was followed by a compensatory mutation in *ntrC*, which reduced the expression of nitrogen uptake and metabolism genes further up-regulated flagellar and chemotaxis gene expression to WT levels (Taylor et al., 2015).

Growth rates and translation

Substrate sensing and induction of the appropriate metabolic pathways is fundamental to successful establishment on the plant hosts and growth is inherently tied into transcription and translation. Alterations in the physiochemical environment impact directly on ribosome structure and function and therefore have a direct impact on effective translation. Functioning bacterial ribosomes are the 70S particles that comprise two subunits, 30S and 50S. In *E. coli*, the small subunit, 30S, is made of 16S rRNA (1,542 nt) and 21 ribosomal proteins (r-proteins), while the large subunit, 50S, is composed of two rRNAs, 23S (2,904 nt) and 5S (120 nt) rRNA, and 33 proteins (Kaczanowska and Rydén-Aulin, 2007). RNA chaperones, helicases and ribosome-dependent GTPases play a role in assisting with RNA folding, by resolving and destabilising incorrect RNA structures. As such, some RNA helicases are involved in ribosome biogenesis, e.g. CsdA, a cold-shock inducible ATP-independent RNA helicase, is a member of the DEAD-box family and is involved in 50S biogenesis (Kaczanowska and Rydén-Aulin, 2007). This protein was found to be induced in *E. coli* O157:H7 (accession # ECs4043) on exposure to spinach leaf lysates when the bacteria were able to grow, but was repressed in a spinach root exudate extract that did not support growth (Crozier et al., 2016). It had been established that under these conditions, the bacterial cultures did not experience any temperature shifts, including a cold shock, but the shift from minimal synthetic medium to plant extracts did result in induction of some of the cold-stress genes. This was indicative of translational stalling by the ribosomes are the metabolic pathways were re-routed to adjust to the different metabolites and explains a role for CsdA. Furthermore, RNA chaperones that are linked to translation inhibition, CspA and CspG, were induced, coincident with induction of *spoT*, a marker of the stringent response (Crozier et al., 2016). A similar adjustment was observed for *E. coli* O157:H7 on contact with growing lettuce plants (Linden et al., 2016).

Growth rate correlates with the amount of functional ribosomes, not just the total amount of ribosomes. In stationary phase, excess ribosomes are stored as 100S particles, which are dimers of 70S. Conversion of the 70S ribosomes to the 100S is regarded as a control mechanism, storing unused ribosomes in an inactive form, and also protecting ribosomes from degradation by proteases and nucleases induced in stationary phase (Wada, 1998). The ribosome modulation factor (RMF) plays a key role ribosome function and translation inhibition by RMF occurs partly as a result of loss of amino acyl tRNA binding on conversion to 100S ribosomal particles. Dissociation of 100S and release of RMF to 70S ribosomes regains high translation activity. *rmf* *m*RNA accumulates in the transition from log to stationary phase and continues in stationary phase. However, *rmf* can also be detected during exponential phase in slow-growing cells so that level of expression is inversely related to growth rate (Wada, 1998). This correlation was seen for *E. coli* O157:H7 on exposure to plant extracts that either did or did not support growth, where *rmf* levels were significantly repressed or induced, respectively (Crozier et al., 2016).

Measurement of growth rates of rhizosphere bacteria is challenging and data vary depending on the approach taken, but it fair to say that doubling times are slow in comparison to growth under laboratory conditions. For example, generation times in the order of 106 hours have been reported for *Pseudomonas* species on sugar beet seedlings in sterilised soil (Christensen et al., 1989). Although this data contrasts with data obtained from modelling with ~ 4.6 cells per day for *Pseudomonas* species or ~ 0.6 for *Bacillus* species and one - four cells per day for most species, (Watt et al., 2006), the generation times are substantially greater than seen in animals and are known to be notoriously difficult to measure accurately. A similarly slow generation time was recorded for *E. coli* O157:H7 colonisation of ~ 1.3 day *Nicotiana benthamiana* (leaf apoplast), albeit at a different tissue site from the rhizosphere (Kaths paper), which contrasts with a generation time for the same bacteria in bovine mucus of ~ 3.4 hrs (Bai et al., 2011).

Conclusion and future developments

Detailed molecular analysis of individual systems has elucidated many of the metabolic processes that occur for rhizobacteria. Although the rhizosphere remains a technically challenging habitat to study, this reductionist approach has yielded many pieces of the puzzle to show the dynamic nature of metabolism within the rhizosphere and the flow of metabolites through the system. The research has now reached a stage where these individual pieces can be put together to produce a more coherent picture that shows how different members of the community capable of different metabolic functions combine to generate a complete and functioning rhizosphere, appropriate to any given habitat.

Technological advances in microbiology, such as deep sequencing and *in situ* spectroscopy have opened our eyes to the extraordinary diversity of microbiota associated with the rhizosphere. Although there is a good understanding of gross-level functions, principally in metabolic processes, metagenomics approaches are on the cusp of providing rhizosphere community-level functions in far greater detail. Key areas for future research relate to crop health and productivity, bioremediation of contaminated land, and bio-mining of rhizosphere-derived secondary metabolites and compounds for the pharmaceutical and biotechnology industries.

A promising technological advance is in the application of single cell approaches within a mixed community. Combining Raman spectroscopy with stable isotope labelling has been used to speciate bacteria and determine metabolic functions, for example uncovering a role for an unculturable species (*Acidovorax* sp.) in the degradation of polycyclic aromatic hydrocarbons in groundwater (Huang et al., 2009). Technical limitations that arise from the lack of some carbon-labelled substrates, such as citrate, have been overcome by the addition of labelled water and a reverse-label approach, where the reversion from labelled to unlabelled cells is detected. This has been successfully used to show that in a mixed culture, *E. coli*, which does not encode citrate catabolism genes, can profit from the ability of *Acinetobacter baylyi* to metabolise citrate, and grow (Wang et al., 2016). Although this elegant demonstration was carried out under laboratory conditions, it has the potential to be developed for *in situ* examination in complex environments such as the rhizosphere.

Work on below-ground interactions has always been hampered by technical challenges working with the soil matrix, e.g. the presence of inhibitors that impact any PCR-based examination of rhizosphere bacteria (Holmes et al., 2014), as well as *in situ* work on the roots and rhizosphere. Various techniques have been applied for *in situ* visualisation, with varying degrees of resolution. However, the development of artificial substrates that permit normal root development hold much promise. Transparent soil particles that have the same refractive index of water and have surface ionic exchange properties, allows seamless penetration of light for microscopic examination and have been used to examine *in situ* rhizosphere colonisation for *E. coli* on lettuce seedlings (Downie et al., 2012). An obvious extension of this is for use with mixed rhizosphere communities, labelled with different fluorescent reporters.

Technological developments have brought together mixed disciplines with the aim of better understanding rhizosphere functions and interactions. Future questions, whether on exploitation for crop productivity or in control of pathogens for crop health, will rely on multi-disciplinary teams to continue to uncover hitherto unknown functions for this fascinating habitat.

References

Amoros, I., Alonso, J.L., and Cuesta, G. (2010) *Cryptosporidium* oocysts and *Giardia* cysts on salad products irrigated with contaminated water. *J Food Prot* 73: 1138-1140.

Appanna, V.P., Alhasawi, A.A., Auger, C., Thomas, S.C., and Appanna, V.D. (2016) Phospho-transfer networks and ATP homeostasis in response to an ineffective electron transport chain in *Pseudomonas fluorescens*. *Archives of Biochemistry and Biophysics* 606: 26-33.

Arat, S., Bullerjahn, G.S., and Laubenbacher, R. (2015) A network biology approach to denitrification in *Pseudomonas aeruginosa*. *PLoS ONE* 10: 12.

Badri, D.V., Weir, T.L., van der Lelie, D., and Vivanco, J.M. (2009) Rhizosphere chemical dialogues: plant-microbe interactions. *Curr Opin Biotechnol* 20: 642-650.

Bagley, S.T. (1985) Habitat association of *Klebsiella* species. *Infection Control* 6: 52-58.

Bai, J., McAteer, S.P., Paxton, E., Mahajan, A., Gally, D.L., and Tree, J.J. (2011) Screening of an *E. coli* O157:H7 bacterial artificial chromosome library by comparative genomic hybridization to identify genomic regions contributing to growth in bovine gastrointestinal mucus and epithelial cell colonization. *Front Microbiol* 2: 168.

Bais, H.P., Weir, T.L., Perry, L.G., Gilroy, S., and Vivanco, J.M. (2006) The role of root exudates in rhizosphere interactions with plants and other organisms. *Ann Rev Plant Biol* 57: 233-266.

Bakker, P., Berendsen, R.L., Doornbos, R.F., Wintermans, P.C.A., and Pieterse, C.M.J. (2013) The rhizosphere revisited: root microbiomics. *Front Plant Sci* 4: 7.

Baumler, D.J., Peplinski, R.G., Reed, J.L., Glasner, J.D., and Perna, N.T. (2011) The evolution of metabolic networks of *E. coli*. *BMC Systems Biology* 5: 21.

Bell, K.S., Sebaihia, M., Pritchard, L., Holden, M.T.G., Hyman, L.J., Holeva, M.C. et al. (2004) Genome sequence of the enterobacterial phytopathogen *Erwinia carotovora* subsp. *atroseptica* and characterization of virulence factors. *PNAS* 101: 11105-11110.

Berg, G., Eberl, L., and Hartmann, A. (2005) The rhizosphere as a reservoir for opportunistic human pathogenic bacteria. *Environ Microbiol* 7: 1673-1685.

Bergholz, T.M., Vanaja, S.K., and Whittam, T.S. (2009) Gene expression induced in *Escherichia coli* O157:H7 upon exposure to model apple juice. *Appl Environ Microbiol* 75: 3542-3553.

Brown, C., and Seidler, R.J. (1973) Potential pathogens in the environment: *Klebsiella pneumoniae*, a taxonomic and ecological enigma. *Appl Microbiol* 25: 900-904.

Bulgarelli, D., Schlaeppi, K., Spaepen, S., van Themaat, E.V.L., and Schulze-Lefert, P. (2013) Structure and Functions of the Bacterial Microbiota of Plants. In *Annual Review of Plant Biology, Vol 64*. Merchant, S.S. (ed). Palo Alto: Annual Reviews, pp. 807-838.

Bulgarelli, D., Garrido-Oter, R., Münch, Philipp C., Weiman, A., Dröge, J., Pan, Y. et al. (2015) Structure and function of the bacterial root microbiota in wild and domesticated barley. *Cell Host & Microbe* 17: 392-403.

Burget, E.G., Verma, R., Mølhøj, M., and Reiter, W.-D. (2003) The biosynthesis of L-Arabinose in plants: molecular cloning and characterization of a Golgi-localized UDP-D-xylose 4-epimerase encoded by the MUR4 gene of *Arabidopsis*. *Plant Cell* 15: 523-531.

Chancellor, D.D., Tyagi, S., Bazaco, M.C., Bacvinskas, S., Chancellor, M.B., Dato, V.M., and de Miguel, F. (2006) Green onions: potential mechanism for hepatitis A contamination. *J Food Prot* 69: 1468-1472.

Christensen, H., Funck-Jensen, D., and Kjøller, A. (1989) Growth rate of rhizosphere bacteria measured directly by the tritiated thymidine incorporation technique. *Soil Biology and Biochemistry* 21: 113-117.

Crozier, L., Hedley, P., Morris, J., Wagstaff, C., Andrews, S.C., Toth, I. et al. (2016) Whole-transcriptome analysis of verocytotoxigenic *Escherichia coli* O157:H7 (Sakai) suggests plant-species-specific metabolic responses on exposure to spinach and lettuce extracts. *Front Microbiol* 7.

D'Arrigo, I., Bojanovič, K., Yang, X., Holm Rau, M., and Long, K.S. (2016) Genome-wide mapping of transcription start sites yields novel insights into the primary transcriptome of *Pseudomonas putida*. *Environ Microbiol* 18: 3466-3481.

De Weger, L.A., van der Vlugt, C.I., Wijfjes, A.H., Bakker, P.A., Schippers, B., and Lugtenberg, B. (1987) Flagella of a plant-growth-stimulating *Pseudomonas fluorescens* strain are required for colonization of potato roots. *J Bacteriol* 169: 2769-2773.

Denison, R.F., and Kiers, E.T. (2004) Lifestyle alternatives for rhizobia: mutualism, parasitism, and forgoing symbiosis. *FEMS Microbiol Lett* 237: 187-193.

Desai, T.A., and Rao, C.V. (2010) Regulation of arabinose and xylose metabolism in *Escherichia coli*. *Appl Environ Microbiol* 76: 1524-1532.

Dixon, R., and Kahn, D. (2004) Genetic regulation of biological nitrogen fixation. *Nat Rev Microbiol* 2: 621-631.

Dong, Y., Iniguez, A.L., and Triplett, E.W. (2003a) Quantitative assessments of the host range and strain specificity of endophytic colonization by *Klebsiella pneumoniae* 342. *Plant Soil* 257: 49-59.

Dong, Y., Iniguez, A.L., Ahmer, B.M., and Triplett, E.W. (2003b) Kinetics and strain specificity of rhizosphere and endophytic colonization by enteric bacteria on seedlings of *Medicago sativa* and *Medicago truncatula*. *Appl Environ Microbiol* 69: 1783-1790.

Dos Santos, P.C., Fang, Z., Mason, S.W., Setubal, J.C., and Dixon, R. (2012) Distribution of nitrogen fixation and nitrogenase-like sequences amongst microbial genomes. *BMC Genomics* 13: 162.

Downie, H., Holden, N., Otten, W., Spiers, A.J., Valentine, T.A., and Dupuy, L.X. (2012) Transparent soil for imaging the rhizosphere. *PLoS ONE* 7: e44276.

Einoder-Moreno, M., Lange, H., Grepp, M., Osborg, E., Vainio, K., and Vold, L. (2016) Non-heat-treated frozen raspberries the most likely vehicle of a norovirus outbreak in Oslo, Norway, November 2013. *Epidemiol Infect* 144: 2765-2772.

Etzler ME, M.D. (2009) Viridiplantae. In *Essentials of Glycobiology*. Varki A, C.R., Esko JD (ed). Cold Spring Harbor (NY): Cold Spring Harbor Laboratory Press.

Falomir, M.P., Rico, H., and Gozalbo, D. (2013) *Enterobacter* and *Klebsiella* species isolated from fresh vegetables marketed in Valencia (Spain) and their clinically relevant resistances to chemotherapeutic agents. *Foodborne Pathogen Dis* 10: 1002-1007.

Fink, R.C., Black, E.P., Hou, Z., Sugawara, M., Sadowsky, M.J., and Diez-Gonzalez, F. (2012) Transcriptional responses of *Escherichia coli* K-12 and O157:H7 associated with lettuce leaves. *Appl Environ Microbiol* 78: 1752-1764.

Fones, H.N., and Preston, G.M. (2013) Trade-offs between metal hyperaccumulation and induced disease resistance in metal hyperaccumulator plants. *Plant Pathology* 62: 63-71.

Fouts, D.E., Tyler, H.L., DeBoy, R.T., Daugherty, S., Ren, Q., Badger, J.H. et al. (2008) Complete genome sequence of the N2-fixing broad host range endophyte *Klebsiella pneumoniae* 342 and virulence predictions verified in mice. *PLoS Genet* 4: e1000141.

Francis, I., Holsters, M., and Vereecke, D. (2010) The Gram-positive side of plant-microbe interactions. *Environ Microbiol* 12: 1-12.

Frans J. de Bruijn (2015) *Biological nitrogen fixation.* Wiley Online Library: John Wiley & Sons, Inc.

Freitag, N.E., Port, G.C., and Miner, M.D. (2009) *Listeria monocytogenes* - from saprophyte to intracellular pathogen. *Nat Rev Microbiol* 7: 623-628.

Garcia-Fraile, P., Seaman, J., Karunakaran, R., Edwards, A., Poole, P., and Downie, J.A. (2015) Arabinose and protocatechuate catabolism genes are important for growth of *Rhizobium leguminosarum* biovar *viciae* in the pea rhizosphere. *Plant Soil* 390: 251-264.

Goldfarb, K., Karaoz, U., Hanson, C., Santee, C., Bradford, M., Treseder, K. et al. (2011) Differential growth responses of soil bacterial taxa to carbon substrates of varying chemical recalcitrance. *Front Microbiol* 2.

Grohmann, K., Manthey, J.A., Cameron, R.G., and Buslig, B.S. (1998) Fermentation of galacturonic acid and pectin-rich materials to ethanol by genetically modified strains of *Erwinia*. *Biotechnology Letters* 20: 195-200.

Haahtela, K., Kari, K., and Sundman, V. (1983) Nitrogenase Activity (Acetylene Reduction) of root-associated, cold-climate *Azospirillum*, *Enterobacter*, *Klebsiella*, and *Pseudomonas* species during growth on various carbon sources and at various partial pressures of oxygen. *Appl Environ Microbiol* 45: 563-570.

Haahtela, K., Laakso, T., and Korhonen, T.K. (1986) Associative nitrogen fixation by *Klebsiella* spp.: adhesion sites and inoculation effects on grass roots. *Appl Environ Microbiol* 52: 1074-1079.

Hacquard, S., Garrido-Oter, R., Gonzalez, A., Spaepen, S., Ackermann, G., Lebeis, S. et al. (2015) Microbiota and host nutrition across plant and animal kingdoms. *Cell Host Microbe* 17: 603-616.

Hao, L.Y., Willis, D.K., Andrews-Polymenis, H., McClelland, M., and Barak, J.D. (2012) Requirement of siderophore biosynthesis for plant colonization by *Salmonella enterica*. *Appl Environ Microbiol* 78: 4561-4570.

Harris, S.J., Shih, Y.L., Bentley, S.D., and Salmond, G.P. (1998) The hexA gene of *Erwinia carotovora* encodes a LysR homologue and regulates motility and the expression of multiple virulence determinants. *Mol Microbiol* 28: 705-717.

Hassaninasab, A., Hashimoto, Y., Tomita-Yokotani, K., and Kobayashi, M. (2011) Discovery of the curcumin metabolic pathway involving a unique enzyme in an intestinal microorganism. *PNAS* 108: 6615-6620.

Hayat, R., Ali, S., Amara, U., Khalid, R., and Ahmed, I. (2010) Soil beneficial bacteria and their role in plant growth promotion: a review. *Annals of Microbiology* 60: 579-598.

Holden, N., Pritchard, L., and Toth, I. (2009) Colonization outwith the colon: plants as an alternative environmental reservoir for human pathogenic enterobacteria. *FEMS Microbiol Rev* 33: 689-703.

Holden, N., Jackson, R.W., and Schikora, A. (2015) Plants as alternative hosts for human and animal pathogens. *Front Microbiol* 6.

Holmes, A., Birse, L., Jackson, R.W., and Holden, N. (2014) An optimised method for the extraction of bacterial mRNA from plant roots infected with *Escherichia coli* O157:H7. *Front Microbiol* 5: 286.

Hou, Z., Fink, R.C., Sugawara, M., Diez-Gonzalez, F., and Sadowsky, M.J. (2013) Transcriptional and functional responses of *Escherichia coli* O157:H7 growing in the lettuce rhizoplane. *Food Microbiol* 35: 136–142.

Hou, Z., Fink, R.C., Black, E., Sugawara, M., Zhang, Z., Diez-Gonzalez, F., and Sadowsky, M.J. (2012) Gene expression profiling of *Escherichia coli* in response to interactions with the lettuce rhizosphere. *J Appl Microbiol* 113: 1076–1086.

Huang, W.E., Ferguson, A., Singer, A.C., Lawson, K., Thompson, I.P., Kalin, R.M. et al. (2009) Resolving genetic functions within microbial populations: *in situ* analyses using rRNA and mRNA stable isotope probing coupled with single-cell Raman-fluorescence *in situ* hybridization. *Appl Environ Microbiol* 75: 234-241.

Huang, X.F., Chaparro, J.M., Reardon, K.F., Zhang, R.F., Shen, Q.R., and Vivanco, J.M. (2014) Rhizosphere interactions: root exudates, microbes, and microbial communities. *Botany-Botanique* 92: 9.

Huertas-Rosales, O., Ramos-Gonzalez, M.I., and Espinosa-Urgel, M. (2016) Self-regulation and interplay of Rsm family proteins modulate the lifestyle of *Pseudomonas putida*. *Appl Environ Microbiol* 82: 5673-5686.

Ibekwe, A.M., Grieve, C.M., Papiernik, S.K., and Yang, C.H. (2009) Persistence of *Escherichia coli* O157:H7 on the rhizosphere and phyllosphere of lettuce. *Lett Appl Microbiol* 49: 784-790.

Iniguez, A.L., Dong, Y., and Triplett, E.W. (2004) Nitrogen fixation in wheat provided by *Klebsiella pneumoniae* 342. *Mol Plant Microbe Interact* 17: 1078-1085.

Iniguez, A.L., Dong, Y., Carter, H.D., Ahmer, B.M.M., Stone, J.M., and Triplett, E.W. (2005) Regulation of enteric endophytic bacterial colonization by plant defenses. *Mol Plant Microbe Interact* 18: 169-178.

Jablasone, J., Warriner, K., and Griffiths, M. (2005) Interactions of *Escherichia coli* O157:H7, *Salmonella typhimurium* and *Listeria monocytogenes* plants cultivated in a gnotobiotic system. *Int J Food Microbiol* 99: 7-18.

James, E.K. (2000) Nitrogen fixation in endophytic and associative symbiosis. *Field Crops Research* 65: 197-209.

Jones, J.D.G., and Dangl, J.L. (2006) The plant immune system. *Nature* 444: 323-329.

Jurtshuk P Jr. (1996) Bacterial metabolism. In *Medical Microbiology*. Baron, S. (ed). Galveston (TX): University of Texas Medical Branch at Galveston.

Kaczanowska, M., and Rydén-Aulin, M. (2007) Ribosome biogenesis and the translation process in *Escherichia coli*. *Microbiol Mol Biol Rev* 71: 477-494.

Kisluk, G., and Yaron, S. (2012) Presence and persistence of *Salmonella enterica* serotype Typhimurium in the phyllosphere and rhizosphere of spray-irrigated parsley. *Appl Environ Microbiol* 78: 4030-4036.

Klerks, M.M., Franz, E., van Gent-Pelzer, M., Zijlstra, C., and van Bruggen, A.H. (2007) Differential interaction of *Salmonella enterica* serovars with lettuce cultivars and plant-microbe factors influencing the colonization efficiency. *ISME J* 1: 620-631.

Kotake, T., Yamanashi, Y., Imaizumi, C., and Tsumuraya, Y. (2016) Metabolism of L-arabinose in plants. *Journal of Plant Research* 129: 781-792.

Kroupitski, Y., Golberg, D., Belausov, E., Pinto, R., Swartzberg, D., Granot, D., and Sela, S. (2009) Internalization of *Salmonella enterica* in leaves is induced by light and involves chemotaxis and penetration through open stomata. *Appl Environ Microbiol* 75: 6076-6086.

Kwan, G., Pisithkul, T., Amador-Noguez, D., and Barak, J. (2015) *De novo* amino acid biosynthesis contributes to *Salmonella enterica* growth in alfalfa seedling exudates. *Appl Environ Microbiol* 81: 861-873.

Kyle, J.L., Parker, C.T., Goudeau, D., and Brandl, M.T. (2010) Transcriptome analysis of *Escherichia coli* O157:H7 exposed to lysates of lettuce leaves. *Appl Environ Microbiol* 76: 1375-1387.

Landstorfer, R., Simon, S., Schober, S., Keim, D., Scherer, S., and Neuhaus, K. (2014) Comparison of strand-specific transcriptomes of enterohemorrhagic *Escherichia coli* O157:H7 EDL933 (EHEC) under eleven different environmental conditions including radish sprouts and cattle feces. *BMC Genomics* 15: 353.

Lennon, J.T., Aanderud, Z.T., Lehmkuhl, B.K., and Schoolmaster, D.R. (2012) Mapping the niche space of soil microorganisms using taxonomy and traits. *Ecology* 93: 1867-1879.

Linden, I.V.d., Cottyn, B., Uyttendaele, M., Vlaemynck, G., Heyndrickx, M., Maes, M., and Holden, N. (2016) Microarray-based screening of differentially expressed genes of *E. coli* O157:H7 Sakai during preharvest survival on butterhead lettuce. *Agriculture* 6: 6.

Macarisin, D., Bauchan, G., and Fayer, R. (2010) *Spinacia oleracea* L. leaf stomata harboring *Cryptosporidium parvum* oocysts: a potential threat to food safety. *Appl Environ Microbiol* 76: 555-559.

Made, D., Trubner, K., Neubert, E., Hohne, M., and Johne, R. (2013) Detection and typing of Norovirus from frozen strawberries involved in a large-scale gastroenteritis outbreak in Germany. *Food and Environmental Virology* 5: 162-168.

Mark, G.L., Dow, J.M., Kiely, P.D., Higgins, H., Haynes, J., Baysse, C. et al. (2005) Transcriptome profiling of bacterial responses to root exudates identifies genes involved in microbe-plant interactions. *PNAS* 102: 17454-17459.

Mendes, R., Garbeva, P., and Raaijmakers, J.M. (2013) The rhizosphere microbiome: significance of plant beneficial, plant pathogenic, and human pathogenic microorganisms. *FEMS Microbiol Rev* 37: 634-663.

Mithani, A., Hein, J., and Preston, G.M. (2011) Comparative analysis of metabolic networks provides insight into the evolution of plant pathogenic and nonpathogenic lifestyles in *Pseudomonas*. *Mol Biol Evol* 28: 483-499.

Mohnen, D. (2008) Pectin structure and biosynthesis. *Curr Opin Plant Biol* 11: 266-277.

Monk, J.M., Charusanti, P., Aziz, R.K., Lerman, J.A., Premyodhin, N., Orth, J.D. et al. (2013) Genome-scale metabolic reconstructions of multiple

Escherichia coli strains highlight strain-specific adaptations to nutritional environments. *PNAS*.

Neidhardt, F.C., and Curtiss, R. (1999) *Escherichia coli and Salmonella: cellular and molecular biology*: ASM press Washington, DC:.

Nogales, J., Palsson, B.O., and Thiele, I. (2008) A genome-scale metabolic reconstruction of *Pseudomonas putida* KT2440: iJN746 as a cell factory. *BMC Systems Biology* 2: 20.

Nudler, E., and Mironov, A.S. (2004) The riboswitch control of bacterial metabolism. *Trends in Biochemical Sciences* 29: 11-17.

Okuda, S., Yamada, T., Hamajima, M., Itoh, M., Katayama, T., Bork, P. et al. (2008) KEGG Atlas mapping for global analysis of metabolic pathways. *Nucl Acids Res* 36: W423-W426.

Olsen, J.E., Hoegh-Andersen, K.H., Casadesus, J., and Thomsen, L.E. (2012) The importance of motility and chemotaxis for extra-animal survival of Salmonella enterica serovar Typhimurium and Dublin. *J Appl Microbiol* 113: 560-568.

Paungfoo-Lonhienne, C., Lonhienne, T.G.A., Yeoh, Y.K., Donose, B.C., Webb, R.I., Parsons, J. et al. (2016) Crosstalk between sugarcane and a plant-growth promoting *Burkholderia* species. *Scientific Reports* 6: 14.

Qi, H.S., Li, S.S., Zhao, S.M., Huang, D., Xia, M.L., and Wen, J.P. (2014) Model-driven redox pathway manipulation for improved isobutanol production in *Bacillus subtilis* complemented with experimental validation and metabolic profiling analysis. *PLoS ONE* 9: 10.

Quilliam, R.S., Williams, A.P., and Jones, D.L. (2012) Lettuce cultivar mediates both phyllosphere and rhizosphere activity of *Escherichia coli* O157:H7. *PLoS ONE* 7: e33842.

Rainey, P. (1999) Adaptation of Pseudomonas fluorescens to the plant rhizosphere. *Environ Microbiol* 1: 243 - 257.

Redondo-Nieto, M., Barret, M., Morrissey, J., Germaine, K., Martinez-Granero, F., Barahona, E. et al. (2013) Genome sequence reveals that Pseudomonas fluorescens F113 possesses a large and diverse array of systems for rhizosphere function and host interaction. *BMC Genomics* 14.

Rudrappa, T., and Bais, H.P. (2008) Curcumin, a known phenolic from *Curcuma longa*, attenuates the virulence of *Pseudomonas aeruginosa* PAO1 in whole plant and animal pathogenicity models. *Journal of Agriculture and Food Chemistry* 56: 1955-1962.

Sahu, S.N., Acharya, S., Tuminaro, H., Patel, I., Dudley, K., LeClerc, J.E. et al. (2003) The bacterial adaptive response gene, barA, encodes a novel conserved histidine kinase regulatory switch for adaptation and modulation of metabolism in *Escherichia coli*. *Molecular and Cellular Biochemistry* 253: 167-177.

Santos, C.R., Hoffmam, Z.B., de Matos Martins, V.P., Zanphorlin, L.M., de Paula Assis, L.H., Honorato, R.V. et al. (2014) Molecular mechanisms associated with xylan degradation by *Xanthomonas* plant pathogens. *J Biol Chem* 289: 32186-32200.

Scales, B.S., Dickson, R.P., LiPuma, J.J., and Huffnagle, G.B. (2014) Microbiology, genomics, and clinical significance of the *Pseudomonas fluorescens* species complex, an unappreciated colonizer of humans. *Clinical Microbiology Reviews* 27: 927-948.

Semenov, A.M., Kuprianov, A.A., and van Bruggen, A.H. (2010) Transfer of enteric pathogens to successive habitats as part of microbial cycles. *Microb Ecol* 60: 239-249.

Shen, X., Hu, H., Peng, H., Wang, W., and Zhang, X. (2013) Comparative genomic analysis of four representative plant growth-promoting rhizobacteria in *Pseudomonas*. *BMC Genomics* 14.

Silby, M., Cerdeno-Tarraga, A., Vernikos, G., Giddens, S., Jackson, R., Preston, G. et al. (2009) Genomic and genetic analyses of diversity and plant interactions of *Pseudomonas fluorescens*. *Genome Biol* 10: R51.

Silby, M.W., Winstanley, C., Godfrey, S.A.C., Levy, S.B., and Jackson, R.W. (2011) Pseudomonas genomes: diverse and adaptable. *FEMS Microbiol Rev* 35: 652-680.

Sinha, J., Reyes, S.J., and Gallivan, J.P. (2010) Reprogramming bacteria to seek and destroy an herbicide. *Nat Chem Biol* 6: 464-470.

Smits, T.H.M., and Duffy, B. (2011) Genomics of iron acquisition in the plant pathogen *Erwinia amylovora*: insights in the biosynthetic pathway of the siderophore desferrioxamine E. *Archives of Microbiology* 193: 693-699.

Sourjik, V., and Wingreen, N.S. (2012) Responding to chemical gradients: bacterial chemotaxis. *Curr Opin Cell Biol* 24: 262-268.

Stover, C., Pham, X., Erwin, A., Mizoguchi, S., Warrener, P., Hickey, M. et al. (2000) Complete genome sequence of *Pseudomonas aeruginosa* PA01, an opportunistic pathogen. *Nature* 406: 959 - 964.

Sulochana, M.B., Jayachandra, S.Y., Kumar, S.A., Parameshwar, A.B., Reddy, K.M., and Dayanand, A. (2014) Siderophore as a potential plant growth-promoting agent produced by *Pseudomonas aeruginosa* JAS-25. *Applied Biochemistry and Biotechnology* 174: 297-308.

Swinkels, H.M., Kuo, M., Embree, G., Andonov, A., Henry, B., Buxton, J.A., and Fraser Hlth Environm Hlth Invest, T. (2014) Hepatitis A outbreak in British Columbia, Canada: the roles of established surveillance, consumer loyalty cards and collaboration, February to May 2012. *Eurosurveill* 19: 36-42.

T G Lessie, a., and P V Phibbs, J. (1984) Alternative pathways of carbohydrate utilization in Pseudomonads. *Ann Rev Microbiol* 38: 359-388.

Taguchi, F., Suzuki, T., Inagaki, Y., Toyoda, K., Shiraishi, T., and Ichinose, Y. (2010) The siderophore pyoverdine of *Pseudomonas syringae* pv. *tabaci* 6605 is anilntrinsic virulence factor in host tobacco infection. *J Bacteriol* 192: 117-126.

Taylor, T.B., Mulley, G., Dills, A.H., Alsohim, A.S., McGuffin, L.J., Studholme, D.J. et al. (2015) Evolution. Evolutionary resurrection of flagellar motility via rewiring of the nitrogen regulation system. *Science* 347: 1014-1017.

Toth, I.K., and Birch, P.R. (2005) Rotting softly and stealthily. *Curr Opin Plant Biol* 8: 424-429.

Valverde, C., Lindell, M., Wagner, E.G.H., and Haas, D. (2004) A repeated GGA motif is critical for the activity and stability of the riboregulator RsmY of *Pseudomonas fluorescens*. *J Biol Chem* 279: 25066-25074.

Visioli, G., D'Egidio, S., and Sanangelantoni, A.M. (2015) The bacterial rhizobiome of hyper-accumulators: future perspectives based on omics analysis and advanced microscopy. *Front Plant Sci* 5.

Wada, A. (1998) Growth phase coupled modulation of *Escherichia coli* ribosomes. *Genes to Cells* 3: 203-208.

Wagner, M.R., Lundberg, D.S., del Rio, T.G., Tringe, S.G., Dangl, J.L., and Mitchell-Olds, T. (2016) Host genotype and age shape the leaf and root microbiomes of a wild perennial plant. *Nature Communications* 7: 15.

Walker, T.S., Bais, H.P., Deziel, E., Schweizer, H.P., Rahme, L.G., Fall, R., and Vivanco, J.M. (2004) *Pseudomonas aeruginosa*-plant root interactions. Pathogenicity, biofilm formation, and root exudation. *Plant Physiol* 134: 320-331.

Wang, C., Deng, Z.L., Xie, Z.M., Chu, X.Y., Chang, J.W., Kong, D.X. et al. (2015) Construction of a genome-scale metabolic network of the plant pathogen *Pectobacterium carotovorum* provides new strategies for bactericide discovery. *FEBS Letters* 589: 285-294.

Wang, Y., Song, Y., Tao, Y., Muhamadali, H., Goodacre, R., Zhou, N.-Y. et al. (2016) Reverse and Multiple Stable Isotope Probing to Study Bacterial Metabolism and Interactions at the Single Cell Level. *Analytical Chemistry*.

Watt, M., Silk, W.K., and Passioura, J.B. (2006) Rates of root and organism growth, soil conditions, and temporal and spatial development of the rhizosphere. *Annals of Botany* 97: 839-855.

Webb, B.A., Helm, R.F., and Scharf, B.E. (2016) Contribution of individual chemoreceptors to *Sinorhizobium meliloti* chemotaxis towards amino acids of host and nonhost seed exudates. *Mol Plant Microbe Interact* 29: 231-239.

Wensing, A., Braun, S.D., Buttner, P., Expert, D., Volksch, B., Ullrich, M.S., and Weingart, H. (2010) Impact of siderophore production by *Pseudomonas syringae* pv. *syringae* 22d/93 on epiphytic fitness and biocontrol activity against *Pseudomonas syringae* pv. *glycinea* 1a/96. *Appl Environ Microbiol* 76: 2704-2711.

Wright, K.M., Chapman, S., McGeachy, K., Humphris, S., Campbell, E., Toth, I.K., and Holden, N.J. (2013) The endophytic lifestyle of *Escherichia coli* O157:H7: quantification and internal localization in roots. *Phytopathol* 103: 333-340.

Xu, C., Chen, X., Duan, D., Peng, C., Le, T., and Shi, J. (2015) Effect of heavy-metal-resistant bacteria on enhanced metal uptake and translocation of the Cu-tolerant plant, *Elsholtzia splendens*. *Environmental Science and Pollution Research* 22: 5070-5081.

Yu, X.M., Ai, C.X., Xin, L., and Zhou, G.F. (2011) The siderophore-producing bacterium, *Bacillus subtilis* CAS15, has a biocontrol effect on *Fusarium* wilt and promotes the growth of pepper. *European Journal of Soil Biology* 47: 138-145.

Zere, T.R., Vakulskas, C.A., Leng, Y.Y., Pannuri, A., Potts, A.H., Dias, R. et al. (2015) Genomic targets and features of BarA-UvrY (-SirA) signal transduction systems. *PLoS ONE* 10: 29.

Zhang, X., George, A., Bailey, M., and Rainey, P. (2006) The histidine utilization (*hut*) genes of *Pseudomonas fluorescens* SBW25 are active on plant surfaces, but are not required for competitive colonization of sugar beet seedlings. *Microbiology* 152: 1867 - 1875.

CHAPTER 2

Role of Plasmids in Plant-Bacteria Interactions

Jasper Schierstaedt[1], Nina Bziuk[2], Nemanja Kuzmanović[2], Khald Blau[2], Kornelia Smalla[2] and Sven Jechalke[3]*

[1]Leibniz Institute of Vegetable and Ornamental Crops (IGZ), Department Plant-microbe systems, Theodor-Echtermeyer-Weg 1, 14979 Großbeeren, Germany
[2]Julius Kühn-Institut - Federal Research Centre for Cultivated Plants (JKI), Institute for Epidemiology and Pathogen Diagnostics, Messeweg 11-12, 38104 Braunschweig, Germany
[3]Justus Liebig University Giessen, Institute for Phytopathology, Heinrich-Buff-Ring 26-32, 35392 Gießen, Germany

*sven.jechalke@julius-kuehn.de

DOI: https://doi.org/10.21775/9781912530007.02

Abstract
Plants are colonized by diverse microorganisms, which may positively or negatively influence the plant fitness. The positive impact includes nutrient acquisition-enhancement of resistance to biotic and abiotic stresses, both important factors for plant growth and survival, while plant pathogenic bacteria can cause diseases. Plant pathogens are adapted to negate or evade plant defense mechanisms, e.g. by the injection of effector proteins into the host cells or by avoiding the recognition by the host. Plasmids play an important role in the rapid bacterial adaptation to stresses and changing environmental conditions. In the plant environment, plasmids can further provide a selective advantage for the host bacteria, e.g. by carrying genes encoding metabolic pathways, metal and antibiotic resistances, or pathogenicity-related genes. However, we are only beginning to understand the role of mobile genetic elements and horizontal gene transfer for plant-associated bacteria. In this review, we aim to provide a short update on what is known about plasmids and horizontal gene transfer of plant-associated bacteria and their role in plant-bacteria interactions. Furthermore, we discuss tools available to study the plant-associated mobilome, its transferability, and its bacterial hosts.

Introduction
Plant-associated microorganisms are considered to be of great importance for plant health, plant productivity and ecosystem functioning. They expand the metabolic repertoire of plants, increase the resource uptake and provide novel nutritional and defense pathways (Berendsen et al., 2012; Berg et al., 2014). Therefore, the genetic information provided by the plant microbiome is also called the second genome of the plant (Berendsen et al., 2012). In the phytosphere, mutualistic associations were studied in great detail for rhizobia and mycorrhizae, rhizobacteria with plant growth promoting or biocontrol activity. However, also parasitic interactions with plant pathogens are well-studied today. Plants are able to influence soil properties, e.g. by the release of nutrients and secondary metabolites via root exudation, which are used to combat pathogenic microorganisms while attracting beneficial ones (Badri et al., 2009; Philippot et al., 2013). At the same time, these rhizodeposits (nutrients, exudates, border cells and mucilage) released by the plants to the rhizosphere (soil influenced by the root) are thriving soil microbial growth, density and activity, which are prerequisites for horizontal gene transfer (HGT) (Kroer et al., 1998; Mølbak et al., 2007; Philippot et al., 2013; Pukall et al., 1996; Raaijmakers et al., 2009; van Elsas et al., 2003). The molecular characterization of strains often revealed that the presence of plasmid-encoded products plays a role in the interaction with the plant. Not only the rhizosphere, also the phyllosphere of plants is considered to be conducive to HGT, which can (positively) affect host fitness (van Elsas et al., 2003). Recently, the development and application of tools such as next generation sequencing contributed to understand the role of mobile genetic elements (MGEs) and HGT in the structure, function and evolution of plant-associated bacterial

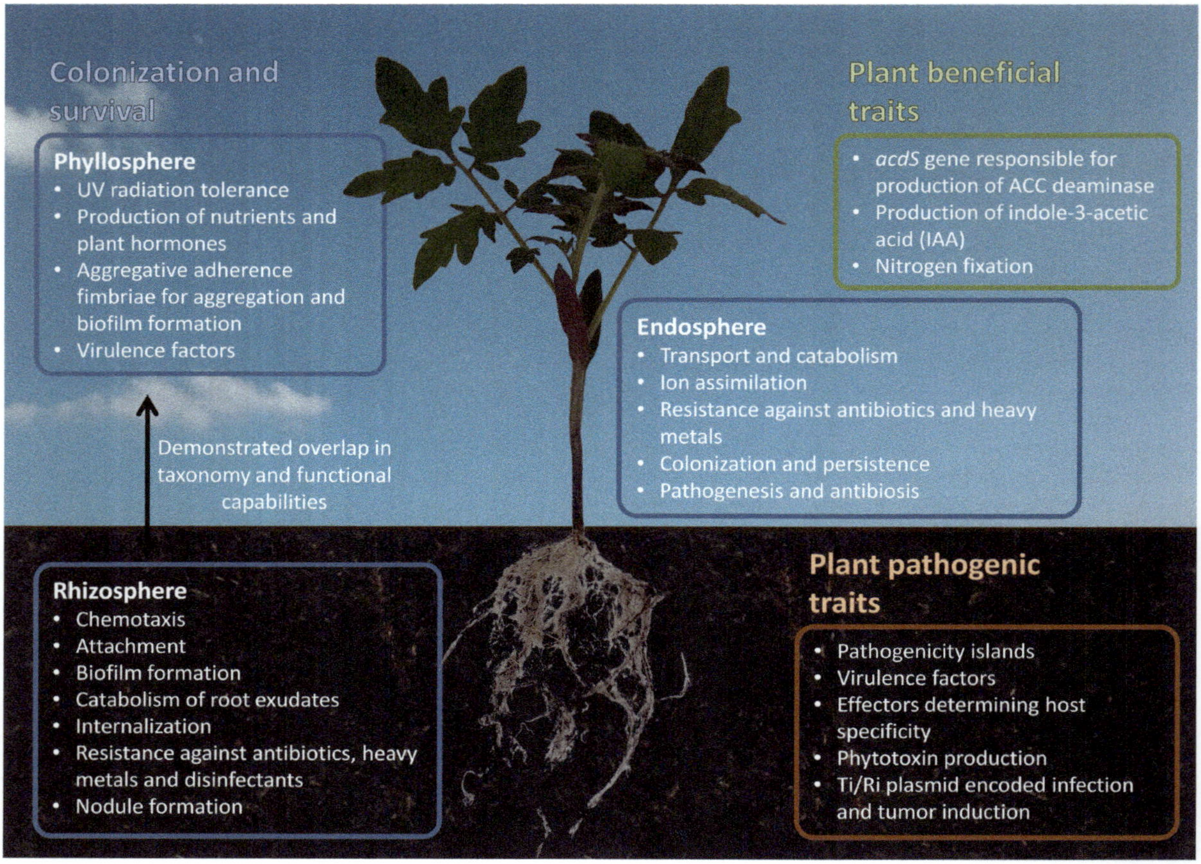

Figure 1. Plasmid-encoded functions in the phytosphere that are described so far. Plasmid-encoded functions are sorted by colonization and survival (blue boxes), plant beneficial (green box) and plant pathogenic traits (brown box). Plant surfaces are considered as hot spots of bacterial conjugation, and an overlap in taxonomy and functional capabilities was already demonstrated between phyllosphere and rhizosphere bacterial communities.

communities in the phytosphere. In this review, we aim to give a short update on plasmids in plant-associated bacteria, HGT and their role in plant-bacteria interactions with a special focus on rhizosphere, phyllosphere and endosphere (summarized in Figure 1). Furthermore, recent methodological developments will be discussed regarding their potential to investigate the plant-associated mobilome and the respective bacterial hosts.

Horizontal gene transfer (HGT) in the phytosphere

HGT has a strong influence on the bacterial evolution (Jain et al., 2002; Koonin et al., 2001; Koski et al., 2001; Ochman et al., 2000; van Elsas et al., 2003). Plasmids belong to the most important vectors for HGT. They provide an efficient mean for rapid bacterial adaption to changing environmental conditions. Additionally to the core (backbone) genes that include plasmid replication, maintenance and transfer, plasmids typically carry a flexible (accessory) gene pool (Heuer and Smalla, 2012). Flexible genes carried by plasmids are known to code for detoxification, virulence, ecological interactions and antibiotic resistance (Smillie et al., 2010), but can also include catabolic pathways (Dennis, 2005). Recombination with the host chromosome and with other plasmids can lead to an acquisition or loss of these functions, resulting in a mosaic and modular genetic composition (Norman et al., 2009; Toussaint and Merlin, 2002). It was reported that a high proportion of bacteria isolated from the phytosphere carry plasmids that are characterized by a high diversity in terms of function and genetic relatedness (Viegas et al., 1997). Furthermore, the phytosphere is constituted by habitats differing with respect to environmental conditions and surface characteristics that require specific adaptations of the colonizing bacteria and provide diverse conditions for HGT. Therefore, in the following chapters, plasmids in bacteria colonizing the different plant habitats, namely rhizosphere, phyllosphere and endosphere are separately discussed in terms of presence, transferability and potential function.

Plant-bacteria interactions in the rhizosphere and the role of plasmids

Root exudates are shaping the rhizosphere bacterial community

The rhizosphere is defined as the soil that is surrounding the root and influenced by the plant (Hartmann et al., 2008a; Hiltner, 1904). In this environment, the plant is in tight contact with soil-borne microorganisms, which include beneficial, saprophytic and pathogenic bacteria, all having a great impact on plant growth and health (Berendsen et al., 2012; Hayat et al., 2010). Plants are able to shape the rhizosphere bacterial community by releasing a wide range of so-called root exudates, which include mono- and polysaccharides, amino acids, sterols, phenols, enzymes, plant growth regulators and different other secondary metabolites (Bais et al., 2006). These compounds form gradients in soil, which attract motile bacteria chemotactically and select for a specific bacterial community (Badri et al., 2009; Bais et al., 2006; Philippot et al., 2013). Accordingly, it was shown for soil bacteria that most genomes contain chemotaxis and mobility genes, providing a competitive advantage in the colonization of plant root surfaces and leading to an enrichment of bacteria carrying chemotaxis-encoding genes in the rhizosphere compared to bulk soil (Scharf et al., 2016). These genes associated with motility but also with adhesion and biofilm formation can be located on plasmids, as described, for example, for pCSA2 of *Cronobacter sakazakii* (Choi et al., 2015). Another indication for the importance of plasmids regarding the rhizosphere competence of bacteria was observed for the plant beneficial bacterium *Bacillus amyloliquefaciens* subsp. *plantarum* S499 that carries a plasmid with seven genes coding for traits predicted to be important for the colonization of plant roots (Molinatto et al., 2016).

In the rhizosphere, the abundance of micro-organisms is higher than in the surrounding bulk soil. However, the evenness is decreased, likely because of the higher availability of carbon and other nutrients released by plant roots (Hartmann et al., 2008b; Kandeler et al., 2002). For example, a considerable amount of the carbon produced by photosynthesis is released by the roots (Marschner, 1995). The microbes able to utilize these nutrients can proliferate. Therefore, the acquisition and exchange of additional metabolic pathways by HGT might be a successful strategy. For example, the self-transmissible plasmid pRme41a, which has been isolated from *Ensifer meliloti* 41, codes for catabolism of root exudates which might be important for the competitiveness in the rhizosphere (Tepfer et al., 1988). Wang et al. (2007) isolated bacteria from the rhizosphere of *Zea mays,* which carried several plasmids conferring the ability to degrade phenol and depicted the connection to polluted sites. Lilley and Bailey (1997) linked fitness advantage of bacteria in the rhizosphere to the acquisition and carriage of plasmids. Furthermore, the cryptic gene-mobilizing plasmid pIPO2 isolated from wheat rhizosphere was found to be a highly proficient IncQ plasmid mobilizer (van Elsas et al., 1998). This plasmid was the first isolated from the PromA group, which is a typical plasmid group of the rhizosphere (Van der Auwera et al., 2009). The whole plasmid was sequenced and its prevalence in soil was assessed (Tauch et al., 2002).

In another study, Jechalke et al. (2014b) observed an enrichment of *korB* genes specific for IncP-1 plasmids in total community (TC)-DNA extracted from the rhizosphere of lettuce compared to bulk soil. However, the isolation and characterization of IncP-1 plasmids from the rhizosphere would be required to gain insights into the traits conferred. In conclusion, plasmids encoding chemotaxis, mobility, colonization and metabolic pathway-associated traits might provide a fitness advantage to their bacterial hosts colonizing the rhizosphere leading to proliferation of plasmid-containing bacteria. Higher plasmid transfer frequencies in the rhizosphere compared to bulk soil are assumed to be a result of the increased abundance of plasmid donors and also of their increased metabolic activity (Pukall et al., 1996).

Soil pollutants

The rhizosphere is a highly dynamic and heterogenous habitat that is shaped by many biotic and abiotic factors. Among these factors are man-made pollutants such as antibiotics, heavy metals, disinfectants and pesticides that can enter the soil e.g. via sewage sludge, digestates, manure or other organic fertilizers (Heuer et al., 2012; Schlüter et al., 2007; Slater et al., 2008; Top et al., 1995; Wolters et al., 2016), and are assumed to increase the abundance of plasmid carrying bacteria. Soil and plant-associated bacteria can adapt to, and cope with these pollutants by means of genes frequently located on MGEs such as plasmids. Bacteria carrying plasmids might be indigenous soil bacteria or introduced to soil, e.g. with organic fertilizers (Binh et al., 2008; Dröge et al., 1999; Hall et al., 2015; Heuer et al., 2012; Jechalke et al., 2014a; Lilley and Bailey, 1997; Malik et al., 2008). Plasmid-mediated traits can provide a fitness advantage for their hosts in case of selective pressure. This was demonstrated for a sulfonamide resistance-carrying LowGC-type plasmid in *Acinetobacter baylyi* BD413 applied to soil treated with manure spiked or unspiked with the veterinary antibiotic sulfadiazine (Jechalke et al., 2013b). Narrow host range (NHR) plasmids, such as the LowGC plasmids, were frequently captured by exogenous isolation from manured soil, maize or grass rhizospheres (Kopmann et al., 2013) and *Acinetobacter* spp. were identified as potential hosts, suggesting their importance for human medicine (Heuer et al., 2009).

Besides antibiotic resistance, plasmids can shuttle other useful traits like heavy metal resistance genes, efflux pumps and toxin-antitoxin systems (reviewed by Heuer and Smalla (2012)). These might provide fitness advantages for their host in the rhizosphere. Accordingly, correlations between the abundance of plasmids and pollution in soils were proposed in numerous studies (de Lipthay et al., 2008; Dealtry et al., 2014; Gstalder et al., 2003; Heuer et al., 2009; Heuer and Smalla, 2012; Smalla et al., 2006; Top et al., 1995), supporting the hypothesis that plasmids are very important for bacteria to cope with, or benefit from soil pollutants in the rhizosphere.

Phyllosphere bacterial communities and their plasmids

The phyllosphere is defined as the aerial surface of the plants. Therefore, tolerance to UV exposure, nutrient and water limitations, as well as high temperature shifts and the presence of reactive oxygen species (ROS) are properties required to successfully colonize the phyllosphere (Knief et al., 2011; Lindow and Brandl, 2003; Newton et al., 2010). Phyllosphere-associated bacteria are supposed to play an important role in plant host growth, fitness and productivity, e.g. by production of growth-promoting nutrients and hormones, protection against pathogen infection, and indirectly influence plant population dynamics and community diversity (Kembel et al., 2014; Laforest-Lapointe et al., 2016; Laforest-Lapointe et al., 2017). Nutrients provided by the leaf can attract bacteria by chemotaxis, as already discussed for the rhizosphere. Furthermore, it was demonstrated that light-treatment of iceberg lettuce leaves resulted in aggregation of *Salmonella enterica* serovar Typhimurium near open stomata and invasion into the inner leaf tissue (Kroupitski et al., 2009). Attachment of bacteria to plant cells is considered to be the next step required in plant bacteria interactions and the formation of biofilms (Rodríguez-Navarro et al., 2007), which can be also associated with MGEs. For example, plasmid-encoded aggregative adherence fimbriae I (AAF/I) of *E. coli* O104:H4 were shown to play a crucial role in aggregation and biofilm formation on spinach and abiotic surfaces (Nagy et al., 2016). Genes linked to type IV secretion systems and extracellular polymeric substance (EPS) synthesis were also detected by Schmeisser et al. (2009) on the megaplasmid pNGR234*b* of *Rhizobium* sp. strain NGR234. Members of the pPT23A-family were reported to likely contribute to virulence and ecological fitness in *Pseudomonas syringae* pathosystems, e.g. by encoding methyl-accepting chemotaxis proteins or conferring UV radiation tolerance (Cazorla et al., 2008; Sundin et al., 2004). Furthermore, bacteria of the genus *Pseudomonas* are frequently found in association with plants and can have significant impact on agriculture by acting as mutualists, saprophytes or pathogens (Espinosa-Urgel, 2004).

Plant surfaces in general are regarded as hotspots of bacterial conjugation events (van Elsas et al., 2003). Considering that the estimated global leaf area is larger than the plain surface of the planet and colonized on average by 10^6-10^7 bacterial cells/cm^2 (Lindow and Brandl, 2003; Ortega et al., 2016; Woodward and Lomas, 2004), the phyllosphere has an enormous potential for bacterial evolution and adaptation by HGT processes. However, the knowledge of plasmid transfer in and on plants is still scarce compared to other environments. Several reports documented high rates of plasmid transfer in the phyllosphere. For example, Normander et al. (1998) examined the transfer of TOL plasmids on the phytoplane of bean (*Phaseolus vulgaris*) and found the plasmid from the donor *Pseudomonas putida* KT2442 in 33% of *P. putida* KT2440 recipients. The transfer ratios on leaves were reported to be 30 times higher than on polycarbonate filters (Normander et al., 1998). Conjugal plasmid transfer between two *P. putida* strains on the phylloplane of bean was positively related to the spatial distribution of the bacteria into microhabitats such as junctures between epidermal cells and substomatal cavities (Normander et al., 1998). Furthermore, the authors concluded that leaf exudates ensured that the activity of the bacteria was sufficient to allow conjugative transfer.

In another study, mercury resistance plasmids were acquired by conjugation from natural bacterial communities of sugar beet roots and leaves to a *Pseudomonas fluorescens* recipient, previously isolated from sugar beets (Lilley and Bailey, 1997), demonstrating that gene transfer within the plant surface compartments might be a common event.

Plasmid-mediated resistome associated with produce

The high potential for HGT in the phyllosphere plays an important role in dissemination of antibiotic resistance genes (ARGs). Conjugal transfer of plasmid-mediated antibiotic resistance from pseudomonads to other *Pseudomonas* spp., Enterobacteriaceae members and to diverse indigenous bacteria occurs on growing alfalfa sprouts (Mølbak *et al.*, 2003) and on the leaves of bean plants (Björklöf et al., 1995). However, produce can be a source of *Escherichia coli* and other Enterobacteriaceae carrying a diverse set of ARGs located on MGEs (Ben Said et al., 2015; Jones-Dias et al., 2016). Conjugative plasmids isolated from produce were shown to carry resistances against antibiotics and against heavy metal compounds or disinfectants, making co-selection possible (Ben Said et al., 2015; Kim et al., 2015; Verraes et al., 2013). Previous studies reported that Enterobacteriaceae and *Pseudomonas* spp., isolated from raw vegetables harbored NHR plasmid groups like IncF, IncI, IncY, and IncX and broad-host-range (BHR) plasmid groups such as IncP, IncH, IncQ IncM/L. These replicon types can carry genes conferring resistance to different antibiotic classes including aminoglycosides, beta-lactams, phenicols, tetracyclines, sulfonamides, and quinolones (Araújo et al., 2017; Ben Said et al., 2015; Jones-Dias et al., 2016; Rahube et al., 2014; Veldman et al., 2014). In addition, the IncF plasmids can also carry virulence genes (Johnson and Nolan, 2009; Villa et al., 2010). Recently, Enterobacteriaceae were isolated from vegetables, soil and irrigation water, which harbored IncI1, IncF, IncK, IncY, and IncB/O plasmids that were reported to carry extended-spectrum beta-lactamase (ESBL) genes (*bla* $_{CTX-M}$) (Ben Said et al., 2015; Kim et al., 2015; Njage and Buys, 2015; Reuland et al., 2014). Mellmann et al. (2011) reported that the NHR IncI1 plasmids carrying *bla* $_{CTX-M-15}$ and *bla* $_{TEM-1}$ have been associated with an outbreak of *E. coli* O104 in Germany in 2011. Interestingly, *E. coli* and *Raoultella omithinolytica* strains recovered from lettuce were found to carry IncHI2, IncI2, and IncX4 plasmids coding for colistin resistance (*mcr-1*) (Luo et al., 2017). Furthermore, plasmid-mediated quinolone resistance (PMQR) genes (*qnrB*, *qnrS*, and *aac(6')-1b-cr*) have been reported in Enterobacteriaceae isolated from fresh herbs, located on IncHI1, IncHI2, IncR, and IncF plasmids (Veldman et al., 2014). Thus, the NHR and BHR plasmids have the potential to be major contributors to the proliferation of ARGs and could pose a potential risk to human health by consumption of produce. However, these studies neglected HGT processes in the plant endosphere, those will be discussed in the following paragraph.

Bacterial plant endophytes

Endophytes are defined as organisms that live or at least persist inside of plants without visibly harming them (Hallmann et al., 1997; Hardoim et al., 2015). Recent technological advances in the field of microbiology and "omics" methods provided fascinating insights into the functional characteristics of endophytes and endophytic communities. In many cases, the coexistence of plant and bacteria had a positive influence on plant growth and the tolerance towards different stresses (Compant et al., 2010; Miliute et al., 2015; Santoyo et al., 2016). The internal root microbiome is typically made up of a community of endophytes interacting and competing with each other (Gaiero et al., 2013). Although not much is known so far about the role of plasmids in endophyte bacterial communities, plasmids might help endophytes to successfully survive, compete within the endophyte community and persist within the plant. In this respect, a cryptic plasmid from an endophytic *Pantoea agglomerans* was isolated from eucalyptus and four potential ORFs were identified with unknown function (de Lima Procópio et al., 2011). In another work, large *Pantoea* plasmids (LPP-1) of 20 *Pantoea* strains including pathogens and endo-/epiphytes were compared, it was revealed that these plasmids contribute to important cell functions like thiamine biosynthesis and played a major role in the adaptation to their ecological niches and functional specialization due to variable elements with genes for the transport and catabolism of diverse metabolic substrates (De Maayer et al., 2012). The small cryptic plasmid pLK39 isolated from a leaf endophytic *Salmonella* sp. carried putative mobilization genes and a replication control region with a high similarity to plasmids isolated from plant pathogenic *Erwinia* species and further ORFs with unknown functions (Dourado et al., 2014), further indicating a potential importance of plasmids for plant colonization. HGT among plant-associated endophytic bacteria might be used by the endophytic bacterial community to adapt to changing environmental conditions. This was demonstrated by Taghavi et al. (2005), who observed the transfer of a toluene degradation pathway located on plasmid pTOM-Bu61 from *Burkholderia cepacia* strains to members of the indigenous plant bacterial communities of poplar. These examples indicate that HGT processes within endophyte bacterial communities are of importance for bacterial adaptation and survival within the host plant.

Plasmids of plant beneficial bacteria

Probably one of the best investigated examples of beneficial plant-bacteria relationships is the symbiosis of leguminous plants with bacteria belonging to the genus *Rhizobium*. The nodule inducing rhizobia are attracted by flavonoids released by the plant resulting in the activation of the plasmid-encoded *nod* operon (Downie, 2010; Venturi and Keel, 2016). The specific expression of genes in both partners results in cellular differentiation processes leading to nodule development and bacterial invasion, finally allowing the bacteria to fix atmospheric nitrogen that can be utilized by the plant (D'Haeze and Holsters, 2002). Nodulation genes of rhizobia, which are involved in the production of Nod factors are primarily located on large symbiotic plasmids, also referred to as Sym plasmids or pSyms (Bánfalvi et al., 1981; Rosenberg et al., 1981). In rhizobia, a distinction can be made between two types of plasmids. Quorum sensing (QS)-regulated plasmids can be transferred by conjugation when the cell density is high enough. Genes encoding the QS regulatory system can be found on the rhizobial plasmids pRL1jl, pRleVF39, pRetCFN42 and pSmed (Ding and Hynes, 2009). Furthermore, some plasmids code for the genes *rctA/B*. While RctA suppresses the transcription of *virB* by binding to the promoter region, RctB is able to antagonize the effect. The *rctA/B* genes can be found on the plasmids pAtC, pRetCFN, pretCIAT and pSymA (Ding and Hynes, 2009). The different plasmid transfer systems of rhizobia were reviewed by Ding and Hynes (2009) suggesting that plasmids play an important role in rhizobia.

In the plant growth-promoting rhizobacterium *Azospirillum brasilense,* plasmids were suggested to mediate bacteria-root-interactions by carrying genes encoding lipopolysaccharides (LPS), EPS and polar and lateral flagella (Fibach-Paldi et al., 2012; Vanbleu et al., 2004). Many plant beneficial soil bacteria were found to produce plant hormones and some bacteria have 1-aminocyclopropane-1-carboxylic acid (ACC)-deaminase genes (Dimkpa et al., 2009; Glick, 2012). For example, *Pseudomonas fluorescens* FY32 was reported to carry a plasmid of 50 kb which codes for the ACC-deaminase gene *acdS* (Farajzadeh et al., 2010). Phylogenetic analysis of ACC-deaminase genes indicated HGT potential of this plasmid (Hontzeas et al., 2005). Genes encoding ACC-deaminase were reported to occur on the chromosome but also on plasmids (Nascimento et al., 2014). ACC-deaminase is able to reduce the ethylene level in plants under stress conditions by cleaving ACC, the immediate precursor of ethylene (Gontia-Mishra et al., 2014). The plant phytohormone auxin plays also a role in the plant's defense system against phytopathogenic bacteria (Spaepen and Vanderleyden, 2011). One of the major naturally occurring hormones of the auxin class, Indole-3-acetic acid (IAA) was found to be plasmid-encoded in wheat root associated *Acinetobacter* strains (Huddedar et al., 2002). This plant hormone is known to affect plant growth and immune response, and it can be produced and used by bacteria as a signaling molecule (Spaepen et al., 2007). Bacteria can use IAA to circumvent the plant defense by de-repressing the auxin signaling in the plant. IAA can also be used as a signaling molecule by bacteria, inducing the expression of genes related to survival under stress conditions (Remans et al., 2006). Localization of genes on high copy plasmids can be used by bacteria to upregulate the expression of genes that are important for plant-bacteria interactions. For example, the localization of auxin biosynthesis genes on multicopy plasmids can lead to a higher IAA production (Spaepen and Vanderleyden, 2011).

Besides hormones there are different other mechanisms of protecting the plant like the secretion of EPS and the induction of plant heat shock proteins and osmo-protectants (Grover et al., 2011). Many EPS-related genes were found to be localized on plasmids in different bacteria (Schmid et al., 2015). Plasmid carriage of EPS-related genes might be advantageous due to the copy number effects.

Plasmids of plant pathogenic bacteria

Plant pathogenic bacteria are responsible for extensive losses in agriculture. Plasmids are generally widely distributed in plant pathogenic bacteria and may carry traits relevant for virulence and ecological fitness of its hosts. Also the host specificity of the pathogen *P. syringae* is determined by effectors that are encoded on plasmids (Bever et al., 2012). A large native plasmid detected in the bean pathogen *P. syringae* pv. *phaseolicola* was found to carry genes encoding type III effectors and other potential virulence genes (Jackson et al., 1999). Furthermore, production of chlorosis-inducing phytotoxin coronatine and an ethylene-forming enzyme, were found to be encoded by plasmids in some pathovars (Alarcón-Chaidez et al., 1999; Bender et al., 1991; Weingart et al., 1999). For the plant pathogenic bacterium *Xanthomonas campestris* pv. *campestris*, it was suggested that the trans-acting sRNA Xcc1 was originally captured by integrons from natural environments and then spread among bacterial species by HGT, mediated by transposons and plasmids (Chen et al., 2011). *X. campestris* can cause black rot disease in cruciferous crops, but it can also be found in asymptomatic association with plant tissue, or as epiphyte (Maas et al., 1985). Marques et al. (2001) analyzed the genome of the plant pathogenic bacterium *Xylella fastidiosa* and revealed two plasmids. In the analysis of the genes present on the plasmids, some ORFs were found that might be important for virulence, and one ORF showed similarity to the virulence-associated protein VapD. *Erwinia* spp. plasmids were found to carry traits such as streptomycin resistance genes, but also pathogenicity and virulence factors such as EPS production and type III secretion systems (T3SSs), which might have caused the emergence of pathogenic from the non-pathogenic species by HGT (Llop, 2015). Furthermore, *Pseudomonas savastanoi*, the causal agent of olive knot disease, harbors plasmids that affect the virulence and plant host range (Bardaji et al., 2011; Eltlbany et al., 2012; Pérez-Martínez et al., 2008). Pathogenicity of *Pantoea agglomerans* causing tumors on various plants is mainly encoded by a plasmid designated as pPATH (Manulis and Barash, 2003). This plasmid carries various virulence genes, including *hrp* cluster and genes encoding type III virulence effectors and biosynthesis of phytohormones. It was also demonstrated that pathogenicity of most *Rhodococcus* spp., causal agents of leafy gall disease on a wide range of host plants, may be associated with a linear plasmid (Creason et al., 2014).

In terms of pathogenesis, tumorigenic and rhizogenic *Rhizobiaceae* species (here collectively called agrobacteria) are certainly the most extensively studied group of phytopathogenic bacteria interacting with plants in the rhizosphere through plasmid-encoded functions. Tumorigenic and rhizogenic Rhizobiaceae strains causing crown gall and hairy root diseases, respectively, are predominantly soil-inhabiting bacteria that are able to colonize rhizosphere and different plant tissues (Bosmans et al., 2017; Burr et al., 1998; Burr and Otten, 1999; Escobar and Dandekar, 2003; Kuzmanović et al., 2018; Otten et al., 2008; Puławska, 2010). They may also systemically colonize host plants and can be present in asymptomatic plant tissue (Otten et al., 2008). Numerous plants can be infected by these bacteria, including important agricultural crops such as fruit species, grapevine and perennial ornamentals. Symptoms of crown gall include tumor formation on roots, crowns, trunks and canes of infected plants. On the other hand, hairy root disease is characterized by abnormal rooting of the host plant. Naturally occurring pathogenic strains are generally distributed within the genus *Agrobacterium*, and species *Allorhizobium vitis* and *Rhizobium rhizogenes*. In the next section, the role of plasmids in plant-bacteria interaction and pathogenesis of bacteria associated with crown gall and hairy root diseases will be discussed in more detail.

Plasmids in bacteria associated with crown gall and hairy root diseases

Bacteria associated with crown gall and hairy root diseases may carry a variable number of plasmids. However, their ability to cause neoplastic diseases is mainly encoded by tumor-inducing (Ti) and root-inducing (Ri) plasmids, carried by tumorigenic and rhizogenic strains, respectively. In addition, plasmids can highly determine the host range of agrobacteria, as shown for the Ti plasmid (Knauf et al., 1982; Thomashow et al., 1980). It was shown that *Agrobacterium fabrum* C58 strains lacking the Ti plasmid did not bind to plant protoplasts, which is likely associated with the virulence induced type IV secretion systems (T4SSs) (Aguilar et al., 2011; Matthysse et al., 1978). The size of Ti/Ri plasmids can range from around 170 kbp to more than 250 kbp. The Ti and Ri plasmids are genetically related and belong to the *repABC* family of megaplasmids (Christie and Gordon, 2014; Pappas and Cevallos, 2011; Suzuki et al., 2009). The Ti/Ri plasmids consists of several functional elements including: transferred DNA (T-DNA), virulence (*vir*) region, opine utilization genes, replication (*rep*) region, and conjugative transfer genes (*tra* and *trb* loci). T-DNA contains genes that are expressed in the host plant and they may be divided into two groups: oncogenes that direct synthesis of phytohormones (Britton et al., 2008), and gene clusters involved in synthesis of special classes of compounds, typically conjugates of amino acids and α-ketoacids or sugars, called opines (Chilton et al., 2001; Dessaux et al., 1998).

Wounds caused by abiotic or biotic factors serve as an entry points for agrobacteria and are considered to be necessary for the initiation of infection, which primarily occurs in the rhizosphere. Interestingly, transformation of unwounded plants by agrobacteria has also been demonstrated (Brencic et al., 2005). Although transformed plants produced opines, tumor formation was absent. The infection itself is a complex process of interkingdom DNA transfer and represents an example of natural genetic transformation of plants. Mechanisms of pathogenesis induced by agrobacteria, including the history of this research field, have been comprehensively reviewed (Gelvin, 2012; Hooykaas, 2000; Kado, 2014; McCullen and Binns, 2006; Nester, 2014; Pitzschke and Hirt, 2010; Zhu et al., 2000; Zupan et al., 2000).

Opines produced by transformed plants, mostly in tumors or hairy roots, serve as a selective nutrient source for the disease-causing rhizobia. The presence of opines is not limited to tumor tissue, they can be translocated to the other parts of the plant and excreted from their roots. Therefore, they may affect also the composition of the bacterial community in the rhizosphere (Mansouri et al., 2002; Oger et al., 1997; Savka et al., 1996; Savka and Farrand, 1997). Moreover, some opines induce the conjugative transfer of Ti plasmids, which is regulated by the QS system belonging to the LuxR/LuxI class (Dessaux et al., 1998; Farrand, 1998; Lang and Faure, 2014; White and Winans, 2007). Agrobacteria may carry some other plasmids that are associated with plant-microbe interactions, which can also specifically interact with Ti/Ri plasmids (Otten et al., 2008; Platt et al., 2014). Thus, non-pathogenic and pathogenic Rhizobiaceae strains isolated from tumors or soil around diseased plants may carry opine-catabolic (OC) plasmids, which contain genes encoding uptake and catabolism of opines as Ti/Ri plasmids (Merlo and Nester, 1977; Petit et al., 1983; Puławska et al., 2016; Szegedi et al., 1999; Wabiko et al., 1990; Wetzel et al., 2014). Unlike Ti/Ri plasmids, OC plasmids do not carry *vir* genes and T-DNA. However, non-pathogenic strains harboring OC plasmids could benefit from opines produced in tumors and in hairy roots. *A. vitis* associated with grapevine crown gall may harbor another type of plasmids, responsible for utilization of tartrate. These plasmids most likely enhance competitiveness of this pathogen on grapevine, since tartrate is an abundant compound in this plant species (Kado, 1998; Salomone et al., 1998; Szegedi et al., 1992). The well-known biological control strain *R. rhizogenes* K84 shows antimicrobial activity against a certain spectrum of agrobacteria, and it was shown to carry the plasmids pAgK84 and pAtK84a that are associated with the synthesis of the specific antibiotics agrocin 84 and 434, respectively (Donner et al., 1993; Ellis et al., 1979; Kim et al., 2006; McClure et al., 1998). Additionally, this strain carries an OC plasmid (pAtK84b) conferring competitiveness of this strain within tumors. Moreover, studies on interaction between Ti plasmid (pTiC58) and accessory megaplasmid (pAtC58) coexisting in *Agrobacterium* sp. C58, suggested that pAtC58 may have a positive effect on tumor size (Nair et al., 2003) and that different forms of this plasmid which arose by deletions can affect *vir* gene expression in pTiC58 (Morton et al., 2013). It was also demonstrated that this plasmid provides bacteria with competitive advantage in the rhizosphere (Morton et al., 2014). Interestingly, some strains of *Agrobacterium* spp. and *R. rhizogenes* may harbor symbiotic (Sym) plasmids and effectively colonize legume plants (Cummings et al., 2009; Velázquez et al., 2005; Zhao et al., 2014).

Tools to investigate plasmids and horizontal gene transfer in the phytosphere

Cultivation-dependent and -independent methods to investigate plasmids

Methods available for plasmid detection and characterization were recently reviewed by Smalla et al. (2015). The vast majority of studies on plasmids of phytosphere-associated bacteria is still based on isolates and thus limited to the fraction of cultivable bacteria. The big advantage of studies on isolates is that the ecology of the plasmid's host and the plasmid's function can be assessed. The presence of plasmids was often revealed as a result of genome sequencing projects as in the case of the citrus pathogen *Xylella fastidiosa* (Marques et al., 2001). However, it is known that only a rather small proportion of environmental bacteria are accessible to cultivation. Thus, to gain a more complete picture, cultivation-independent methods are increasingly used to detect and quantify plasmid abundance in the phytosphere (Smalla et al., 2015). By extracting the TC-DNA from the phytosphere, the plasmids in the whole bacterial population present in a representative sample type can be studied. However, plasmids typically are present only in a small fraction of the bacterial community and in low abundance (Heuer et al., 2008). Thus, to detect plasmids in TC-DNA by PCR-amplification using primers for conserved regions, Southern blot hybridization is required to increase the sensitivity and specificity of plasmid detection (Dealtry et al., 2014). Quantitative real-time PCR allows the quantification of plasmids in a microbial community, and its relative abundance is provided by relating the plasmid-specific gene copy numbers to the 16S rRNA gene copy number, as it was done with *trfA* and *korB* for IncP-1 plasmids (Heuer et al., 2012; Jechalke et al., 2013a). To our knowledge, recently developed tools to study plasmid occurrence and diversity by directly sequencing the plasmidome (Li et al., 2012) or by determining the potential plasmid host by EPIC-PCR (Spencer et al., 2015) were not yet employed for phytosphere-associated bacteria. The diversity of IncP-1 plasmids was recently studied by amplicon sequencing of the *trfA* gene (Dealtry et al., 2014), but this approach was not yet used to study the IncP-1 plasmid diversity in the phytosphere.

Additionally, transferable plasmids can be captured independently from the cultivability of their original hosts. Different plasmid-capture approaches are known, e.g. the *in vitro* transposon-aided capture method (TRACA). An EZ-*Tn5 OriV* Kan2 transposon is inserted (Jones and Marchesi, 2006), and no selectable markers, conjugative function or mobilization are needed. Interestingly, plasmids captured by means of exogenous plasmid isolations via biparental or triparental mating from the rhizosphere of a wide a range of crops such as wheat (van Elsas et al., 1998), alfalfa (Schneiker et al., 2001), maize and grass (Jechalke et al., 2013c), but also from the mycosphere (Zhang et al., 2014) were often affiliated to BHR plasmids of the IncP-1 or the pPromA group. There is no doubt that the progress made in next generation sequencing technologies strongly improved our knowledge of the diversity of plasmids in bacteria colonizing the phytosphere. These tools are of outmost importance to understand the contribution of plasmid-encoded traits to foster the adaptation and diversification of their hosts. A list of selected techniques and their advantages and disadvantages is given in Table 1.

Bioinformatic approaches

Different sequencing techniques are currently available for the characterization of plasmids. With the sequencing by PacBio, comparatively fast sequencing with a very long read length is possible (Rhoads and Au, 2015). The long reads facilitate the plasmid sequence assembly. Even though the PacBio method is hindered by a relatively low throughput and high error rate, this sequencing approach might be complemented with the Illumina approach. This combination enabled the sequencing of the natural plasmid pMM259 that mediates genetic exchange by HGT between rhizobia and *Roseobacter* species (Bartling et al., 2017). Furthermore, whole genome sequencing by PacBio revealed a plasmid in the biocontrol strain *Bacillus amyloliquefaciens* subsp. Plantarum S499 (Molinatto et al., 2016), which was later assumed to be important for root colonization and control of surfactant production, as well as biofilm formation (Molinatto et al., 2017). Plant pathogenic bacteria were studied by PacBio as well, e.g. Lu et al. (2018) revealed the presence of pathogenicity-related plasmids in *Clavibacter michiganensis* subsp. *insidiosus* which can cause crop diseases. The megaplasmid found in *Ralstonia solanacearum* contains genes for type II and III secretion systems (Li et al., 2018). Another rather new sequencing possibility is the nanopore sequencing technology, which enables the sequencing of single DNA molecules. The available technologies are reviewed in more detail in Goodwin et al. (2016).

Nowadays, many bioinformatics tools are available for different plasmid analyses. The first step in the analysis is the distinction between chromosomal and plasmid DNA. Plasmids can be identified using tools such as cBar, plasmidSPAdes, PlasmidFinder or MLST (Arredondo-Alonso et al., 2016; Carattoli et al., 2014; Zhou and Xu, 2010). The reads can be mapped to a reference database using SRST2 (Inouye et al., 2014). Plasmid reconstruction methods are very useful to analyze plasmids by clustering dendrograms, allowing the comparison of plasmid diversity and adaptation (de Toro et al., 2014). Examples for available tools include PLACNET, Recycler and plasmidSPAdes (Antipov et al., 2016; Lanza et al., 2014; Rozov et al., 2016). Reconstructed plasmids can be investigated for potential functions, such as the presence of resistance genes (Clausen et al., 2016; Gupta et al., 2014; McArthur et al., 2013; Zankari et al., 2012). Several of the applications were reviewed in Orlek et al. (2017) and Edwards and Holt (2013).

Table 1: Advantages and disadvantages of selected tools for the investigation of plasmids in the phytosphere.

Methods	Advantages	Disadvantages
Plating of environmental bacteria followed by plasmid extraction	○ Easy to perform ○ Focus on markers possible by selective plating	○ Limited to the cultivable fraction of bacteria ○ Not all environmental bacteria are suitable for plasmid extraction
PCR-amplification of known backbone genes from total community DNA	○ Represents the whole community ○ Combined with Southern blot hybridization it allows a more sensitive analysis	○ Original plasmid host remains unknown ○ Limited detection of novel sequences
Quantification of plasmids relative to 16S gene copy number by quantitative real time PCR	○ Determination of plasmid abundance, e.g. extracted from total community DNA ○ More specific than PCR e.g.by use of additional TaqMan probe	○ Original plasmid host remains unknown ○ More expensive than PCR
Pyrosequencing	○ Detection of unknown plasmids/sequences, even when the host is unknown	○ Difficult assembly ○ Not state-of-the-art method
Exogenous plasmid isolation by bi- or triparental mating	○ Independent from original host and its cultivability ○ Possibility to capture novel plasmids	○ Original plasmid host remains unknown ○ Depends on selectable markers ○ Depends on replication in recipient and transferability of plasmid
Transposon aided capture method (TRACA)	○ Culture-independent ○ Does not rely on plasmid-encoded traits ○ Possibility to capture novel plasmids	○ Not all captured plasmids are stable in *E. coli* ○ Large plasmids are difficult to capture due to low copy number and challenging transformation
GFP-tagged plasmids	○ Easy quantification and localization of the plasmids ○ Offers potential to discover original hosts/new hosts/ transfer rate etc	○ Might be difficult to transfer ○ Potential metabolic burden
Illumina sequencing	○ Rapidly decreasing costs ○ High throughput	○ Short read length
PacBio sequencing	○ Very long read length	○ Low throughput
Nanopore sequencing	○ Single molecule sequencing ○ Potential of long read length ○ Low cost	○ Single-base resolution potentially reduced by stochastic motion of the DNA molecule

Plasmid transfer, maintenance and evolution studies

The transfer and distribution of plasmids can also be investigated *in situ* via incorporation of marker genes. In this approach, *gfp*-tagged plasmids under the control of a *lac* promoter are used together with donor bacteria that have a chromosomally inserted *laclq* repressor. While there is no *gfp* transcription in the donor cell, once the plasmid enters the recipient bacterium, the *lac* promoter is not under the control of the repressor anymore and the *gfp* is expressed (Klümper et al., 2015). By fluorescent microscopy or flow cytometry, plasmid transfer can be quantified and transconjugants isolated. Such approach was already applied for the rhizosphere community of barley (Musovic et al., 2006). Labeling of plasmids with marker genes like *gfp* allows the investigate of their transfer and also their stability. Plasmid transfer can be investigated through growth of the recipient on selective media (Bale et al., 1988). However, this method is limited to the cultivable fraction of a bacterial community.

Most of the studies on the cost and benefits of plasmids for their hosts were done in rich nutrient broth. However, these studies might provide only little insights into plasmid cost and benefits for bacteria residing in natural ecosystems. Although it is well documented that a subset of soil bacteria is enriched in the phytosphere, the generation times are assumed to be much longer and thus plasmid carriage might be less costly. Recently, Bziuk et al. (unpublished) examined the potential fitness cost or benefit of the IncP-1 plasmid pTL25 that was captured by triparental mating from the rhizosphere of lettuce grown in non-polluted soil. A competition experiment was performed by inoculating lettuce plants grown in soil microcosms with plasmid-carrying and plasmid-free *Pseudomonas putida* KT2442 (50:50 ratio). While a stable maintenance of pTL25 carrying *P. putida* KT2442 was observed in the rhizosphere, its proportion in bulk soil decreased. These findings indicate that in the rhizosphere carriage of plasmid pTL25 seemed not to be disadvantageous.

Plasmid evolutionary studies might answer the question of the contribution of some plasmids to the versatility of plant-associated bacteria. Plasmid evolution occurs mainly through cost amelioration and stability improvement (Hughes et al., 2012). These adaptations of plasmids to their host were described previously (Bouma and Lenski, 1988). Nevertheless, evolutionary studies in the phytosphere are still scarce.

Conclusions

Over the past decades an increasing number of studies on plasmids in the phytosphere increased our knowledge of their diversity and functions in the plant environment. In general, plasmids in plant associated bacteria facilitates their survival in the phytosphere environment, with the rapidly changing conditions and diverse biotic and abiotic stresses. Plasmid-encoded traits can contribute to an improvement of plant growth and health. However, an increased ecological understanding of plasmids, related functions and their respective hosts is needed for the development of new crop breeding strategies. Hence, plasmid-encoded functions could help the plant to better respond to biotic and abiotic stresses.

In addition, plasmids are also important carriers of plant-pathogenic traits. To improve disease prevention, it is essential to investigate their function, virulence and fitness contribution, as well as their host specificity. While the microbiome of different crops is presently studied in detail, systematic studies on their plasmidome were not yet undertaken. Furthermore, the presence of pollutants such as pesticides, veterinary medicines, metal compound introduced via organic fertilizers or irrigation water can foster the abundance of bacteria carrying plasmids with antibiotic or heavy metal resistance genes. In particular, the role of co-selection of antibiotic resistance by heavy metal or disinfectant compounds needs further investigation. The hidden diversity of plasmids in the rare phytosphere microbiome can only be assessed by a polyphasic approach comprising isolates, the molecular analysis of TC-DNA from detached cells or from enrichments and exogenously captured plasmids. These approaches would allow to detect antibiotic resistance plasmids in enteric bacteria that are typically not very abundant in the phytosphere such as *E. coli* or *Salmonella*. These might be taken up by humans via raw fruits and vegetables and present risks for human health.

Rapid developments in next generation sequencing techniques offer the ability to sequence plasmids from plant-associated bacteria or plasmids captured into recipients. However, due to their low abundance, the direct sequencing of plasmids extracted from the phytosphere is still difficult. Nevertheless, a large number of plasmid sequences are already available. These allow comparative studies on plasmids and the development of tools for systematic studies on the occurrence and diversity in the phytosphere. However, it remains a challenging but very important task to investigate the cost and benefit of plasmids for their host bacteria in the phytosphere in order to understand the role of plasmids in interactions between bacteria and plants.

Acknowledgements
The work of Sven Jechalke and Jasper Schierstaedt was supported by the Federal Office for Agriculture and Food (Bundesanstalt für Landwirtschaft und Ernährung, BLE), Grant 13HS026 and 13HS029. The work of Nina Bziuk was supported by the Federal Ministry of Education and Research (Bundesministerium für Bildung und Forschung, BMBF), Grant 031B0196B. Nemanja Kuzmanović was supported by the Georg Forster Fellowship for postdoctoral researchers from the Alexander von Humboldt-Foundation, Bonn, Germany. Work of Khald Blau was supported by a scholarship from the Libyan government.

References

Alarcón-Chaidez, F.J., Peñaloza-Vázquez, A., Ullrich, M., and Bender, C.L. (1999). Characterization of plasmids encoding the phytotoxin coronatine in *Pseudomonas syringae*. Plasmid *42*, 210-220.

Antipov, D., Hartwick, N., Shen, M., Raiko, M., Lapidus, A., and Pevzner, P.A. (2016). plasmidSPAdes: assembling plasmids from whole genome sequencing data. Bioinformatics *32*, 3380-3387.

Araújo, S., A.T. Silva, I., Tacão, M., Patinha, C., Alves, A., and Henriques, I. (2017). Characterization of antibiotic resistant and pathogenic *Escherichia coli* in irrigation water and vegetables in household farms. Int J Food Microbiol *257*, 192-200.

Arredondo-Alonso, S., van Schaik, W., Willems, R.J., and Schurch, A.C. (2016). On the (im)possibility to reconstruct plasmids from whole genome short-read sequencing data. bioRxiv.

Badri, D.V., Weir, T.L., van der Lelie, D., and Vivanco, J.M. (2009). Rhizosphere chemical dialogues: plant-microbe interactions. Curr Opin Biotechnol *20*, 642-650.

Bais, H.P., Weir, T.L., Perry, L.G., Gilroy, S., and Vivanco, J.M. (2006). The role of root exudates in rhizosphere interactions with plants and other organisms. Annu Rev Plant Biol *57*, 233-266.

Bale, M.J., J., D.M., and Fry, J.C. (1988). Novel method for studying plasmid transfer in undistrubed river epilithon. Appl Environ Microbiol *54*, 2756-2758.

Bánfalvi, Z., Sakanyan, V., Koncz, C., Kiss, A., Dusha, I., and Kondorosi, A. (1981). Location of nodulation and nitrogen fixation genes on a high molecular weight plasmid of *R. meliloti*. Mol Gen Genet *184*, 318-325.

Bardaji, L., Pérez-Martínez, I., Rodríguez-Moreno, L., Rodríguez-Palenzuela, P., Sundin, G.W., Ramos, C., and Murillo, J. (2011). Sequence and role in virulence of the three plasmid complement of the model tumor-inducing bacterium *Pseudomonas savastanoi* pv. savastanoi NCPPB 3335. PLoS One *6*, e25705.

Bartling, P., Brinkmann, H., Bunk, B., Overmann, J., Göker, M., and Petersen, J. (2017). The composite 259-kb plasmid of *Martelella mediterranea* DSM 17316T-a natural replicon with functional RepABC modules from Rhodobacteraceae and Rhizobiaceae. Front Microbiol *8*.

Ben Said, L., Jouini, A., Klibi, N., Dziri, R., Alonso, C.A., Boudabous, A., Ben Slama, K., and Torres, C. (2015). Detection of extended-spectrum beta-lactamase (ESBL)-producing Enterobacteriaceae in vegetables, soil and water of the farm environment in Tunisia. Int J Food Microbiol *203*, 86-92.

Bender, C.L., Young, S.A., and Mitchell, R.E. (1991). Conservation of plasmid DNA sequences in coronatine-producing pathovars of *Pseudomonas syringae*. Appl Environ Microbiol *57*, 993-999.

Berendsen, R.L., Pieterse, C.M., and Bakker, P.A. (2012). The rhizosphere microbiome and plant health. Trends Plant Sci *17*, 478-486.

Berg, G., Grube, M., Schloter, M., and Smalla, K. (2014). The plant microbiome and its importance for plant and human health. Front Microbiol *5*.

Bever, J.D., Platt, T.G., and Morton, E.R. (2012). Microbial population and community dynamics on plant roots and their feedbacks on plant communities. Annu Rev Microbiol *66*, 265-283.

Binh, C.T.T., Heuer, H., Kaupenjohann, M., and Smalla, K. (2008). Piggery manure used for soil fertilization is a reservoir for transferable antibiotic resistance plasmids. FEMS Microbiol Ecol *66*, 25-37.

Björklöf, K., Suoniemi, A., Haahtela, K., and Romantschuk, M. (1995). High frequency of conjugation versus plasmid segregation of RP1 in epiphytic *Pseudomonas syringae* populations. Microbiology *141*, 2719-2727.

Bosmans, L., Moerkens, R., Wittemans, L., De Mot, R., Rediers, H., and Lievens, B. (2017). Rhizogenic agrobacteria in hydroponic crops: epidemics, diagnostics and control. Plant Pathol *66*, 1043 - 1053.

Bouma, J.E., and Lenski, R.E. (1988). Evolution of a bacteria/plasmid association. Nature *335*, 351 - 352.

Brencic, A., Angert, E.R., and Winans, S.C. (2005). Unwounded plants elicit *Agrobacterium vir* gene induction and T-DNA transfer: Transformed plant cells produce opines yet are tumour free. Molecular Microbiology *57*, 1522-1531.

Britton, M.T., Escobar, M.A., and Dandekar, A.M. (2008). The oncogenes of *Agrobacterium tumefaciens* and *Agrobacterium rhizogenes*. In Agrobacterium: From Biology to Biotechnology, T. Tzfira, and V. Citovsky, eds. (New York, NY: Springer New York), pp. 523-563.

Burr, T.J., Bazzi, C., Süle, S., and Otten, L. (1998). Crown gall of grape: Biology of *Agrobacterium vitis* and the development of disease control strategies. Plant Dis *82*, 1288-1297.

Burr, T.J., and Otten, L. (1999). Crown gall of grape: Biology and disease management. Annu Rev Phytopathol *37*, 53-80.

Carattoli, A., Zankari, E., García-Fernández, A., Larsen, M.V., Lund, O., Villa, L., Aarestrup, F.M., and Hasman, H. (2014). In silico detection and typing of plasmids using PlasmidFinder and plasmid multilocus sequence typing. Antimicrob Agents Chemother *58*, 3895 - 3903.

Cazorla, F.M., Codina, J.C., Abad, C., Arrebola, E., Torés, J.A., Murillo, J., Pérez-Garcia, A., and de Vicente, A. (2008). 62-kb plasmids harboring *rulAB* homologues confer UV-tolerance and epiphytic fitness to *Pseudomonas syringae* pv. *syringae* mango isolates. Microb Ecol *56*, 283-291.

Chen, X.-L., Tang, D.-J., Jiang, R.-P., He, Y.-Q., Jiang, B.-L., Lu, G.-T., and Tang, J.-L. (2011). sRNA-Xcc1, an integron-encoded transposon- and plasmid-transferred trans-acting sRNA, is under the positive control of the key virulence regulators HrpG and HrpX of *Xanthomonas campestris* pathovar *campestris*. RNA Biol *8*, 947-953.

Chilton, W.S., Petit, A., Chilton, M.D., and Dessaux, Y. (2001). Structure and characterization of the crown gall opines heliopine, vitopine and ridéopine. Phytochemistry *58*, 137-142.

Choi, Y., Kim, S., Hwang, H., Kim, K.P., Kang, D.H., and Ryu, S. (2015). Plasmid-encoded MCP is involved in virulence, motility, and biofilm formation of *Cronobacter sakazakii* ATCC 29544. Infect Immun *83*, 197-204.

Christie, P.J., and Gordon, J.E. (2014). The *Agrobacterium* Ti Plasmids. Microbiol Spectr *2*, 10.1128/microbiolspec.PLAS-0010-2013.

Clausen, P.T.L.C., Zankari, E., Aarestrup, F.M., and Lund, O. (2016). Benchmarking of methods for identification of antimicrobial resistance genes in bacterial whole genome data. J Antimicrob Chemother *71*, 2484-2488.

Compant, S., Clément, C., and Sessitsch, A. (2010). Plant growth-promoting bacteria in the rhizo- and endosphere of plants: Their role, colonization, mechanisms involved and prospects for utilization. Soil Biol Biochem *42*, 669-678.

Creason, A.L., Vandeputte, O.M., Savory, E.A., Davis, E.W., II, Putnam, M.L., Hu, E., Swader-Hines, D., Mol, A., Baucher, M., Prinsen, E., *et al.* (2014). Analysis of genome sequences from plant pathogenic *Rhodococcus* reveals genetic novelties in virulence loci. PLoS One *9*, e101996.

Cummings, S.P., Gyaneshwar, P., Vinuesa, P., Farruggia, F.T., Andrews, M., Humphry, D., Elliott, G.N., Nelson, A., Orr, C., Pettitt, D., *et al.* (2009). Nodulation of *Sesbania* species by *Rhizobium* (*Agrobacterium*) strain IRBG74 and other rhizobia. Environ Microbiol *11*, 2510-2525.

D'Haeze, W., and Holsters, M. (2002). Nod factor structures, responses, and perception during initiation of nodule development. Glycobiology *12*, 79R-105R.

de Lima Procópio, R.E., Araújo, W.L., Andreote, F.D., and Azevedo, J.L. (2011). Characterization of a small cryptic plasmid from endophytic *Pantoea agglomerans* and its use in the construction of an expression vector. Genet Mol Biol *34*, 103-109.

de Lipthay, J.R., Rasmussen, L.D., Oregaard, G., Simonsen, K., Bahl, M.I., Kroer, N., and Sørensen, S.J. (2008). Acclimation of subsurface microbial communities to mercury. FEMS Microbiol Ecol *65*, 145-155.

De Maayer, P., Chan, W.-Y., Blom, J., Venter, S.N., Duffy, B., Smits, T.H.M., and Coutinho, T.A. (2012). The large universal *Pantoea* plasmid LPP-1 plays a major role in biological and ecological diversification. BMC Genomics *13*, 625.

de Toro, M., Garcilláon-Barcia, M.P., and La Cruz, F.d. (2014). Plasmid diversity and adaptation analyzed by massive sequencing of *Escherichia coli* plasmids. Microbiol Spectr *2*.

Dealtry, S., Ding, G.C., Weichelt, V., Dunon, V., Schlüter, A., Martini, M.C., Del Papa, M.F., Lagares, A., Amos, G.C.A., Wellington, E.M.H., *et al.* (2014). Cultivation-independent screening revealed hot spots of IncP-1, IncP-7 and IncP-9 plasmid occurrence in different environmental habitats. PLoS One *9*, e89922.

Dennis, J.J. (2005). The evolution of IncP catabolic plasmids. Curr Opin Biotechnol *16*, 291-298.

Dessaux, Y., Petit, A., Farrand, S.K., and Murphy, P.J. (1998). Opines and opine-like molecules involved in plant-*Rhizobiaceae* interactions. In The *Rhizobiaceae*, Molecular Biology of Model Plant-Associated Bacteria, H.P. Spaink, A. Kondorosi, and P.J.J. Hooykaas, eds. (Dordrecht, The Netherlands: Kluwer Academic Publishers), pp. 173-197.

Dimkpa, C., Weinand, T., and Asch, F. (2009). Plant-rhizobacteria interactions alleviate abiotic stress conditions. Plant Cell Environ *32*, 1682-1694.

Ding, H., and Hynes, M.F. (2009). Plasmid transfer systems in the rhizobia. Can J Microbiol *55*, 917-927.

Donner, S.C., Jones, D.A., McClure, N.C., Rosewarne, G.M., Tate, M.E., Kerr, A., Fajardo, N.N., and Clare, B.G. (1993). Agrocin 434, a new plasmid encoded agrocin from the biocontrol *Agrobacterium* strains K84 and K1026, which inhibits biovar 2 agrobacteria. Physiol Mol Plant Pathol *42*, 185-194.

Dourado, P.L., Cardoso, V.L., Coelho, A.S., and Bataus, L.A. (2014). Sequence analysis and characterization of plasmid pLK39 isolated from endophytic *Salmonella* sp. isolated from *Solanum lycocarpum*. Curr Microbiol *69*, 846-851.

Downie, J.A. (2010). The roles of extracellular proteins, polysaccharides and signals in the interactions of rhizobia with legume roots. FEMS Microbiol Rev *34*, 150-170.

Dröge, M., Pühler, A., and Selbitschka, W. (1999). Horizontal gene transfer among bacteria in terrestrial and aquatic habitats as assessed by microcosm and field studies. Biol Fertility Soils *29*, 221-245.

Edwards, D.J., and Holt, K.E. (2013). Beginner's guide to comparative bacterial genome analysis using next-generation sequence data. Microb Inform Exp *3*, 1 - 9.

Ellis, J.G., Kerr, A., Van Montagu, M., and Schell, J. (1979). *Agrobacterium*: genetic studies on agrocin 84 production and the biological control of crown gall. Physiol Mol Plant Pathol *15*, 311-319.

Eltlbany, N., Prokscha, Z.Z., Castañeda-Ojeda, M.P., Krögerrecklenfort, E., Heuer, H., Wohanka, W., Ramos, C., and Smalla, K. (2012). A new bacterial disease on *Mandevilla sanderi*, caused by *Pseudomonas savastanoi*: Lessons learned for bacterial diversity studies. Appl Environ Microbiol *78*, 8492-8497.

Escobar, M.A., and Dandekar, A.M. (2003). *Agrobacterium tumefaciens* as an agent of disease. Trends Plant Sci *8*, 380-386.

Espinosa-Urgel, M. (2004). Plant-associated *Pseudomonas* populations: molecular biology, DNA dynamics, and gene transfer. Plasmid *52*, 139-150.

Farajzadeh, D., Aliasgharzad, N., Bashir, N.S., and Yakhchali, B. (2010). Cloning and characterization of a plasmid encoded ACC deaminase from an indigenous *Pseudomonas fluorescens* FY32. Curr Microbiol *61*, 37-43.

Farrand, S.K. (1998). Conjugal plasmids and their transfer. In The Rhizobiaceae: Molecular Biology of Model Plant-Associated Bacteria, H.P. Spaink, A. Kondorosi, and P.J.J. Hooykaas, eds. (Dordrecht: Springer Netherlands), pp. 199-233.

Fibach-Paldi, S., Burdman, S., and Okon, Y. (2012). Key physiological properties contributing to rhizosphere adaptation and plant growth promotion abilities of *Azospirillum brasilense*. FEMS Microbiol Lett *326*, 99-108.

Gaiero, J.R., McCall, C.A., Thompson, K.A., Day, N.J., Best, A.S., and Dunfield, K.E. (2013). Inside the root microbiome: Bacterial root endophytes and plant growth promotion. Am J Bot *100*, 1738-1750.

Gelvin, S.B. (2012). Traversing the cell: *Agrobacterium* T-DNA's journey to the host genome. Front Plant Sci *3*, 52.

Glick, B.R. (2012). Plant growth-promoting bacteria: Mechanisms and applications. Scientifica *2012*, 1-15.

Gontia-Mishra, I., Sasidharan, S., and Tiwari, S. (2014). Recent developments in use of 1-aminocyclopropane-1-carboxylate (ACC) deaminase for conferring tolerance to biotic and abiotic stress. Biotechnol Lett *36*, 889-898.

Goodwin, S., McPherson, J.D., and McCombie, W.R. (2016). Coming of age: ten years of next-generation sequencing technologies. Nat Rev Genet *17*, 333-351.

Grover, M., Ali, S.Z., Sandhya, V., Rasul, A., and Venkateswarlu, B. (2011). Role of microorganisms in adaptation of agriculture crops to abiotic stresses. World J Microbiol Biotechnol *27*, 1231-1240.

Gstalder, M.-E., Faelen, M., Mine, N., Top, E.M., Mergeay, M., and Couturier, M. (2003). Replication functions of new broad host range plasmids isolated from polluted soils. Res Microbiol *154*, 499-509.

Gupta, S.K., Padmanabhan, B.R., Diene, S.M., Lopez-Rojas, R., Kempf, M., Landraud, L., and Rolain, J.-M. (2014). ARG-ANNOT, a new bioinformatic tool to discover antibiotic resistance genes in bacterial genomes. Antimicrob Agents Chemother *58*, 212 - 220.

Hall, J.P.J., Harrison, E., Lilley, A.K., Paterson, S., Spiers, A.J., and Brockhurst, M.A. (2015).

Environmentally co-occurring mercury resistance plasmids are genetically and phenotypically diverse and confer variable context-dependent fitness effects. Environ Microbiol *17*, 5008-5022.

Hallmann, J., Quadt-Hallmann, A., Mahaffee, W.F., and Kloepper, J.W. (1997). Bacterial endophytes in agricultural crops Can J Microbiol *43*, 895 -914.

Hardoim, P.R., van Overbeek, L.S., Berg, G., Pirttilä, A.M., Compant, S., Campisano, A., Döring, M., and Sessitsch, A. (2015). The hidden world within plants: Ecological and evolutionary considerations for defining functioning of microbial endophytes. Microbiol Mol Biol Rev *79*, 293-320.

Hartmann, A., Rothballer, M., and Schmid, M. (2008a). Lorenz Hiltner, a pioneer in rhizosphere microbial ecology and soil bacteriology research. Plant Soil *312*, 7-14.

Hartmann, A., Schmid, M., Tuinen, D.v., and Berg, G. (2008b). Plant-driven selection of microbes. Plant Soil *321*, 235-257.

Hayat, R., Ali, S., Amara, U., Khalid, R., and Ahmed, I. (2010). Soil beneficial bacteria and their role in plant growth promotion: a review. Ann Microbiol *60*, 579-598.

Heuer, H., Abdo, Z., and Smalla, K. (2008). Patchy distribution of flexible genetic elements in bacterial populations mediates robustness to environmental uncertainty. FEMS Microbiol Ecol *65*, 361-371.

Heuer, H., Binh, C.T.T., Jechalke, S., Kopmann, C., Zimmerling, U., Krögerrecklenfort, E., Ledger, T., González, B., Top, E.M., and Smalla, K. (2012). IncP-1ε plasmids are important vectors of antibiotic resistance genes in agricultural systems: diversification driven by class 1 integron gene cassettes. Front Microbiol *3*, 1 - 8.

Heuer, H., Kopmann, C., Binh, C.T.T., Top, E.M., and Smalla, K. (2009). Spreading antibiotic resistance through spread manure: characteristics of a novel plasmid type with low %G+C content. Environ Microbiol *11*, 937-949.

Heuer, H., and Smalla, K. (2007). Manure and sulfadiazine synergistically increased bacterial antibiotic resistance in soil over at least two months. Environ Microbiol *9*, 657-666.

Heuer, H., and Smalla, K. (2012). Plasmids foster diversification and adaptation of bacterial populations in soil. FEMS Microbiol Rev *36*, 1083-1104.

Hiltner, L. (1904). Über neuere Erfahrungen und Probleme auf dem Gebiet der Bodenbakteriologie und unter besonderer Berücksichtigung der Gründüngung und Brache. Arb DLG *98*, 59-78.

Hontzeas, N., Richardson, A.O., Belimov, A., Safronova, V., Abu-Omar, M.M., and Glick, B.R. (2005). Evidence for horizontal transfer of 1-aminocyclopropane-1-carboxylate deaminase genes. Appl Environ Microbiol *71*, 7556-7558.

Hooykaas, P.J.J. (2000). *Agrobacterium,* a natural metabolic engineer of plants. In Metabolic Engineering of Plant Secondary Metabolism, R. Verpoorte, and A.W. Alfermann, eds. (Dordrecht: Springer Netherlands), pp. 51-67.

Huddedar, S.B., Shete, A.M., Tilekar, J.N., Gore, S.D., Dhavale, D.D., and Chopade, B.A. (2002). Isolation, characterization, and plasmid pUPI126-mediated indole-3-acetic acid production in *Acinetobacter* strains from rhizosphere of wheat. Appl Biochem Biotechnol, pp. 21-39.

Hughes, J.M., Lohman, B.K., Deckert, G.E., Nichols, E.P., Settles, M., Abdo, Z., and Top, E.M. (2012). The role of clonal interference in the evolutionary dynamics of plasmid-host adaptation. mBio *3*, e00077-00012.

Inouye, M., Dashnow, H., Raven, L.-A., Schultz, M.B., Pope, B.J., Tomita, T., Zobel, J., and Holt, K.E. (2014). SRST2: Rapid genomic surveillance for public health and hospital microbiology labs. Genome Med *6*, 90.

Jackson, R.W., Athanassopoulos, E., Tsiamis, G., Mansfield, J.W., Sesma, A., Arnold, D.L., Gibbon, M.J., Murillo, J., Taylor, J.D., and Vivian, A. (1999). Identification of a pathogenicity island, which contains genes for virulence and avirulence, on a large native plasmid in the bean pathogen *Pseudomonas syringae* pathovar *phaseolicola*. Proc Natl Acad Sci USA *96*, 10875-10880.

Jain, R., Rivera, M.C., Moore, J.E., and Lake, J.A. (2002). Horizontal gene transfer in microbial genome evolution. Theor Popul Biol *61*, 489-495.

Jechalke, S., Dealtry, S., Smalla, K., and Heuer, H. (2013a). Quantification of IncP-1 plasmid prevalence in environmental samples. Appl Environ Microbiol *79*, 1410-1413.

Jechalke, S., Heuer, H., Siemens, J., Amelung, W., and Smalla, K. (2014a). Fate and effects of veterinary antibiotics in soil. Trends Microbiol *22*, 536-545.

Jechalke, S., Kopmann, C., Richter, M., Moenickes, S., Heuer, H., and Smalla, K. (2013b). Plasmid-mediated fitness advantage of *Acinetobacter baylyi* in sulfadiazine-polluted soil. FEMS Microbiol Lett *348*, 127-132.

Jechalke, S., Kopmann, C., Rosendahl, I., Groeneweg, J., Weichelt, V., Krögerrecklenfort, E., Brandes, N., Nordwig, M., Ding, G.-C., Siemens, J., *et al.* (2013c). Increased abundance and transferability of resistance genes after field application of manure from sulfadiazine-treated pigs. Appl Environ Microbiol *79*, 1704-1711.

Jechalke, S., Schreiter, S., Wolters, B., Dealtry, S., Heuer, H., and Smalla, K. (2014b). Widespread dissemination of class 1 integron components in soils and related ecosystems as revealed by cultivation-independent analysis. Front Microbiol *4*, 420.

Johnson, T.J., and Nolan, L.K. (2009). Pathogenomics of the virulence plasmids of *Escherichia coli*. Microbiology and Molecular Biology Reviews : MMBR *73*, 750-774.

Jones, B.V., and Marchesi, J.R. (2006). Transposon-aided capture (TRACA) of plasmids resident in the human gut mobile metagenome. Nat Methods *4*, 55-61.

Jones-Dias, D., Manageiro, V., Ferreira, E., Barreiro, P., Vieira, L., Moura, I.B., and Caniça, M. (2016). Architecture of class 1, 2, and 3 integrons from gram negative bacteria recovered among fruits and vegetables. Front Microbiol *7*, 1400.

Kado, C.I. (1998). Origin and evolution of plasmids. Antonie Van Leeuwenhoek *73*, 117-126.

Kado, C.I. (2014). Historical account on gaining insights on the mechanism of crown gall tumorigenesis induced by *Agrobacterium tumefaciens*. Front Microbiol *5*, 340.

Kandeler, E., Marschner, P., Tscherko, D., Singh Gahoonia, T., and Nielsen, N.E. (2002). Microbial community composition and functional diversity in the rhizosphere of maize. Plant Soil *238*, 301-312.

Kembel, S.W., O'Connor, T.K., Arnold, H.K., Hubbell, S.P., Wright, S.J., and Green, J.L. (2014). Relationships between phyllosphere bacterial communities and plant functional traits in a neotropical forest. Proc Natl Acad Sci USA *111*, 13715-13720.

Kim, H.-S., Chon, J.-W., Kim, Y.-J., Kim, D.-H., Kim, M.-s., and Seo, K.-H. (2015). Prevalence and characterization of extended-spectrum-β-lactamase-producing *Escherichia coli* and *Klebsiella pneumoniae* in ready-to-eat vegetables. Int J Food Microbiol *207*, 83-86.

Kim, J.G., Park, B.K., Kim, S.U., Choi, D., Nahm, B.H., Moon, J.S., Reader, J.S., Farrand, S.K., and Hwang, I. (2006). Bases of biocontrol: sequence predicts synthesis and mode of action of agrocin 84, the Trojan horse antibiotic that controls crown gall. Proc Natl Acad Sci USA *103*, 8846-8851.

Klümper, U., Riber, L., Dechesne, A., Sannazzarro, A., Hansen, L.H., Sørensen, S.J., and Smets, B.F. (2015). Broad host range plasmids can invade an unexpectedly diverse fraction of a soil bacterial community. The ISME Journal *9*, 934-945.

Knauf, V.C., Panagopoulos, C.G., and Nester, E.W. (1982). Genetic factors controlling the host range of *Agrobacterium tumefaciens*. Phytopathology *72*, 1545-1549.

Knief, C., Delmotte, N., and Vorholt, J.A. (2011). Bacterial adaptation to life in association with plants - A proteomic perspective from culture to in situ conditions. Proteomics *11*, 3086-3105.

Koonin, E.V., Makarova, K.S., and Aravind, L. (2001). Horizontal gene transfer in prokaryotes: Quantification and classification. Annu Rev Microbiol *55*, 709-742.

Kopmann, C., Jechalke, S., Rosendahl, I., Groeneweg, J., Krögerrecklenfort, E., Zimmerling, U., Weichelt, V., Siemens, J., Amelung, W., Heuer, H., *et al.* (2013). Abundance and transferability of antibiotic resistance as related to the fate of sulfadiazine in maize rhizosphere and bulk soil. FEMS Microbiol Ecol *83*, 125-134.

Koski, L.B., Morton, R.A., and Golding, G.B. (2001). Codon bias and base composition are poor indicators of horizontally transferred genes. Mol Biol Evol *18*, 404-412.

Kroer, N., Barkay, T., Sørensen, S., and Weber, D. (1998). Effect of root exudates and bacterial metabolic activity on conjugal gene transfer in the rhizosphere of a marsh plant. FEMS Microbiol Ecol *25*, 375-384.

Kroupitski, Y., Golberg, D., Belausov, E., Pinto, R., Swartzberg, D., Granot, D., and Sela, S. (2009). Internalization of *Salmonella enterica* in leaves is induced by light and involves chemotaxis and penetration through open stomata. Appl Environmen Microbiol *75*, 6076-6086.

Kuzmanović, N., Puławska, J., Hao, L., and Burr, T.J. (2018). The ecology of *Agrobacterium vitis* and management of crown gall disease in vineyards. Curr Top Microbiol Immunol, doi: 10.1007/82_2018_85.

Laforest-Lapointe, I., Messier, C., and Kembel, S.W. (2016). Tree phyllosphere bacterial communities: exploring the magnitude of intra- and inter-individual variation among host species. PeerJ *4*, e2367.

Laforest-Lapointe, I., Paquette, A., Messier, C., and Kembel, S.W. (2017). Leaf bacterial diversity mediates plant diversity and ecosystem function relationships. Nature *546*, 145-147.

Lang, J., and Faure, D. (2014). Functions and regulation of quorum-sensing in *Agrobacterium tumefaciens*. Front Plant Sci *5*, 14.

Lanza, V.F., de Toro, M., Garcillán-Barcia, M.P., Mora, A., Blanco, J., Coque, T.M., and de la Cruz, F. (2014). Plasmid flux in *Escherichia coli* ST131 sublineages, analyzed by plasmid constellation

network (PLACNET), a new method for plasmid reconstruction from whole genome sequences. PLoS Genet *10*, e1004766.
Li, L.L., Norman, A., Hansen, L.H., and Sørensen, S.J. (2012). Metamobilomics - expanding our knowledge on the pool of plasmid encoded traits in natural environments using high-throughput sequencing. Clin Microbiol Infect *18*, 8-11.
Li, X., Huang, X., Chen, G., Zou, L., Wei, L., and Hua, J. (2018). Complete genome sequence of the sesame pathogen *Ralstonia solanacearum* strain SEPPX 05. Genes and Genomics *40*, 657-668.
Lilley, A.K., and Bailey, M.J. (1997). The acquisition of indigenous plasmids by a genetically marked pseudomonad population colonizing the sugar beet phytosphere is related to local environmental conditions. Appl Environ Microbiol *63*, 1577-1583.
Lindow, S.E., and Brandl, M.T. (2003). Microbiology of the phyllosphere. Appl Environ Microbiol *69*, 1875-1883.
Llop, P. (2015). Genetic islands in pome fruit pathogenic and non-pathogenic *Erwinia* species and related plasmids. Front Microbiol *6*, 8.
Lu, Y., Ishimaru, C.A., Glazebrook, J., and Samac, D.A. (2018). Comparative genomic analyses of *Clavibacter michiganensis* subsp. insidiosus and pathogenicity on *Medicago truncatula*. Phytopathology *108*, 172-185.
Luo, J., Yao, X., Lv, L., Doi, Y., Huang, X., Huang, S., and Liu, J.-H. (2017). Emergence of *mcr-1* in *Raoultella ornithinolytica* and *Escherichia coli* isolates from retail vegetables in China. Antimicrob Agents Chemother *61*, e01139-01117.
Maas, J.L., Finnney, M.M., Civerolo, E.L., and Sasser, M. (1985). Association of an unusual strain of *Xanthomonas campestris* with apple. American Phytopathological Society *75*, 438 - 445.
Malik, A., Çelik, E.-K., Bohn, C., Böckelmann, U., Knobel, K., and Grohmann, E. (2008). Detection of conjugative plasmids and antibiotic resistance genes in anthropogenic soils from Germany and India. FEMS Microbiol Lett *279*, 207-216.
Mansouri, H., Petit, A., Oger, P., and Dessaux, Y. (2002). Engineered rhizosphere: the trophic bias generated by opine-producing plants is independent of the opine type, the soil origin, and the plant species. Appl Environ Microbiol *68*, 2562-2566.
Manulis, S., and Barash, I. (2003). *Pantoea agglomerans* pvs. *gypsophilae* and *betae*, recently evolved pathogens? Molecular Plant Pathology *4*, 307-314.
Marques, M.V., da Silva, A.M., and Gomes, S.L. (2001). Genetic organization of plasmid pXF51 from the plant pathogen *Xylella fastidiosa*. Plasmid *45*, 184-199.
Marschner, H. (1995). The soil-root interface (rhizosphere) in relation to mineral nutrition. In Mineral Nutrition of Higher Plants (Second Edition) (London: Academic Press), pp. 537-595.
McArthur, A.G., Waglechner, N., Nizam, F., Yan, A., Azad, M.A., Baylay, A.J., Bhullar, K., Canova, M.J., De Pascale, G., Ejim, L., *et al.* (2013). The comprehensive antibiotic resistance database. Antimicrob Agents Chemother *57*, 3348-3357.
McClure, N.C., Ahmadi, A.-R., and Clare, B.G. (1998). Construction of a range of derivatives of the biological control strain *Agrobacterium rhizogenes* K84: a study of factors involved in biological control of crown gall disease. Appl Environ Microbiol *64*, 3977-3982.
McCullen, C.A., and Binns, A.N. (2006). *Agrobacterium tumefaciens* and plant cell interactions and activities required for interkingdom macromolecular transfer. Annu Rev Cell Dev Biol *22*, 101-127.
Mellmann, A., Harmsen, D., Cummings, C.A., Zentz, E.B., Leopold, S.R., Rico, A., Prior, K., Szczepanowski, R., Ji, Y., Zhang, W., *et al.* (2011). Prospective genomic characterization of the German enterohemorrhagic *Escherichia coli* O104:H4 outbreak by rapid next generation sequencing technology. PLoS One *6*, e22751.
Merlo, D.J., and Nester, E.W. (1977). Plasmids in avirulent strains of *Agrobacterium*. J Bacteriol *129*, 76-80.
Miliute, I., Buzaite, O., Baniulis, D., and Stanys, V. (2015). Bacterial endophytes in agricultural crops and their role in stress tolerance: a review. Zemdirbyste-Agriculture *102*, 465-478.
Mølbak, L., Molin, S., and Kroer, N. (2007). Root growth and exudate production define the frequency of horizontal plasmid transfer in the rhizosphere. FEMS Microbiol Ecol *59*, 167-176.
Molinatto, G., Franzil, L., Steels, S., Puopolo, G., Pertot, I., and Ongena, M. (2017). Key impact of an uncommon plasmid on *Bacillus amyloliquefaciens* subsp. *plantarum* S499 developmental traits and lipopeptide production. Front Microbiol *8*.
Molinatto, G., Puopolo, G., Sonego, P., Moretto, M., Engelen, K., Viti, C., Ongena, M., and Pertot, I. (2016). Complete genome sequence of *Bacillus amyloliquefaciens* subsp. *plantarum* S499, a rhizobacterium that triggers plant defences and inhibits fungal phytopathogens. J Biotechnol *238*, 56-59.

Morton, E.R., Merritt, P.M., Bever, J.D., and Fuqua, C. (2013). Large deletions in the pAtC58 megaplasmid of *Agrobacterium tumefaciens* can confer reduced carriage cost and increased expression of virulence genes. Genome Biol Evol *5*, 1353-1364.

Morton, E.R., Platt, T.G., Fuqua, C., and Bever, J.D. (2014). Non-additive costs and interactions alter the competitive dynamics of co-occurring ecologically distinct plasmids. Proceedings Biological Sciences / The Royal Society *281*, 20132173.

Musovic, S., Oregaard, G., Kroer, N., and Sørensen, S.J. (2006). Cultivation-independent examination of horizontal transfer and host range of an IncP-1 plasmid among gram-positive and gram-negative bacteria indigenous to the barley rhizosphere. Appl Environ Microbiol *72*, 6687-6692.

Nagy, A., Xu, Y.F., Bauchan, G.R., Shelton, D.R., and Nou, X.W. (2016). Aggregative adherence fimbriae I (AAF/I) mediate colonization of fresh produce and abiotic surface by Shiga toxigenic enteroaggregative *Escherichia coli* O104:H4. Int J Food Microbiol *229*, 44-51.

Nair, G.R., Liu, Z., and Binns, A.N. (2003). Reexamining the role of the accessory plasmid pAtC58 in the virulence of *Agrobacterium tumefaciens* strain C58. Plant Physiology *133*, 989-999.

Nascimento, F.X., Rossi, M.J., Soares, C.R.F.S., McConkey, B.J., and Glick, B.R. (2014). New insights into 1-aminocyclopropane-1-carboxylate (ACC) deaminase phylogeny, evolution and ecological significance. PLoS One *9*, e99168.

Nester, E.W. (2014). *Agrobacterium*: nature's genetic engineer. Front Plant Sci *5*, 730.

Newton, A.C., Gravouil, C., and Fountaine, J.M. (2010). Managing the ecology of foliar pathogens: ecological tolerance in crops. Ann Appl Biol *157*, 343-359.

Njage, P.M.K., and Buys, E.M. (2015). Pathogenic and commensal *Escherichia coli* from irrigation water show potential in transmission of extended spectrum and AmpC β-lactamases determinants to isolates from lettuce. Microbial Biotechnology *8*, 462-473.

Norman, A., Hansen, L.H., and Sørensen, S.J. (2009). Conjugative plasmids: vessels of the communal gene pool. Philos Trans R Soc Lond, Ser B: Biol Sci *364*, 2275-2289.

Normander, B., Christensen, B.B., Molin, S., and Kroer, N. (1998). Effect of bacterial distribution and activity on conjugal gene transfer on the phylloplane of the bush bean (*Phaseolus vulgaris*). Appl Environ Microbiol *64*, 1902-1909.

Ochman, H., Lawrence, J.G., and Groisman, E.A. (2000). Lateral gene transfer and the nature of bacterial innovation. Nature *405*, 299-304.

Oger, P., Petit, A., and Dessaux, Y. (1997). Genetically engineered plants producing opines alter their biological environment. Nat Biotechnol *15*, 369-372.

Orlek, A., Stoesser, N., Anjum, M.F., Doumith, M., Ellington, M.J., Peto, T., Crook, D., Woodford, N., Walker, A.S., Phan, H., *et al.* (2017). Plasmid classification in an era of whole-genome sequencing: Application in studies of antibiotic resistance epidemiology. Front Microbiol *8*.

Ortega, R.A., Mahnert, A., Berg, C., Muller, H., and Berg, G. (2016). The plant is crucial: specific composition and function of the phyllosphere microbiome of indoor ornamentals. FEMS Microbiol Ecol *92*.

Otten, L., Burr, T., and Szegedi, E. (2008). *Agrobacterium*: A disease-causing bacterium. In *Agrobacterium*: From Biology to Biotechnology, T. Tzfira, and V. Citovsky, eds. (New York, USA: Springer), pp. 1-46.

Pappas, K.M., and Cevallos, M.A. (2011). Plasmids of the *Rhizobiaceae* and their role in interbacterial and transkingdom interactions. In: Witzany G. (eds) Biocommunication in Soil Microorganisms. Soil Biology, vol 23. Springer, Berlin, Heidelberg.

Pérez-Martínez, I., Zhao, Y., Murillo, J., Sundin, G.W., and Ramos, C. (2008). Global Genomic Analysis of *Pseudomonas savastanoi* pv. savastanoi Plasmids. J Bacteriol *190*, 625-635.

Petit, A., David, C., Dahl, G.A., Ellis, J.G., Guyon, P., Casse-Delbart, F., and Tempé, J. (1983). Further extension of the opine concept: plasmids in *Agrobacterium rhizogenes* cooperate for opine degradation. Molecular and General Genetics MGG *190*, 204-214.

Philippot, L., Raaijmakers, J.M., Lemanceau, P., and van der Putten, W.H. (2013). Going back to the roots: the microbial ecology of the rhizosphere. Nat Rev Microbiol *11*, 789-799.

Pitzschke, A., and Hirt, H. (2010). New insights into an old story: Agrobacterium-induced tumour formation in plants by plant transformation. EMBO J *29*, 1021-1032.

Platt, T.G., Morton, E.R., Barton, I.S., Bever, J.D., and Fuqua, C. (2014). Ecological dynamics and complex interactions of *Agrobacterium* megaplasmids. Front Plant Sci *5*, 15.

Pukall, R., Tschäpe, H., and Smalla, K. (1996). Monitoring the spread of broad host and narrow host range plasmids in soil microcosms. FEMS Microbiol Ecol *20*, 53-66.

Puławska, J. (2010). Crown gall of stone fruits and nuts, economic significance and diversity of its causal agents: tumorigenic *Agrobacterium* spp. Journal of Plant Pathology *92*, S1.87-S81.98.

Puławska, J., Kuzmanović, N., Willems, A., and Pothier, J.F. (2016). *Pararhizobium polonicum* sp. nov. isolated from tumors on stone fruit rootstocks. Syst Appl Microbiol *39*, 164-169.

Raaijmakers, J.M., Paulitz, T.C., Steinberg, C., Alabouvette, C., and Moënne-Loccoz, Y. (2009). The rhizosphere: A playground and battlefield for soilborne pathogens and beneficial microorganisms. Plant Soil *321*, 341-361.

Rahube, T.O., Marti, R., Scott, A., Tien, Y.-C., Murray, R., Sabourin, L., Zhang, Y., Duenk, P., Lapen, D.R., and Topp, E. (2014). Impact of fertilizing with raw or anaerobically digested sewage sludge on the abundance of antibiotic-resistant coliforms, antibiotic resistance genes, and pathogenic bacteria in soil and on vegetables at harvest. Appl Environ Microbiol *80*, 6898-6907.

Remans, R., Spaepen, S., and Vanderleyden, J. (2006). Auxin signaling in plant defense. Science *313*, 171.

Reuland, E.A., al Naiemi, N., Raadsen, S.A., Savelkoul, P.H.M., Kluytmans, J.A.J.W., and Vandenbroucke-Grauls, C.M.J.E. (2014). Prevalence of ESBL-producing Enterobacteriaceae in raw vegetables. Eur J Clin Microbiol Infect Dis *33*, 1843-1846.

Rhoads, A., and Au, K.F. (2015). PacBio sequencing and its applications. Genomics Proteomics Bioinformatics *13*, 278-289.

Rodríguez-Navarro, D.N., Dardanelli, M.S., and Ruíz-Saínz, J.E. (2007). Attachment of bacteria to the roots of higher plants. FEMS Microbiol Lett *272*, 127-136.

Rosenberg, C., Boistard, P., Denarie, J., and Casse-Delbart, F. (1981). Genes controlling early and late functions in symbiosis are located on a megaplasmid in *Rhizobium meliloti*. Mol Gen Genet *184*, 326-333.

Rozov, R., Brown Kav, A., Bogumil, D., Shterzer, N., Halperin, E., Mizrahi, I., and Shamir, R. (2016). Recycler: An algorithm for detecting plasmids from de novo assembly graphs. bioRxiv.

Salomone, J.Y., Szegedi, E., Cobanov, P., and Otten, L. (1998). Tartrate utilization genes promote growth of *Agrobacterium* spp. on grapevine. Mol Plant-Microbe Interact *11*, 836-838.

Santoyo, G., Moreno-Hagelsieb, G., del Carmen Orozco-Mosqueda, M., and Glick, B.R. (2016). Plant growth-promoting bacterial endophytes. Microbiol Res *183*, 92-99.

Savka, M.A., Black, R.C., Binns, A.N., and Farrand, S.K. (1996). Translocation and exudation of tumor metabolites in crown galled plants. Molecular plant-microbe interactions : MPMI *9*, 310-313.

Savka, M.A., and Farrand, S.K. (1997). Modification of rhizobacterial populations by engineering bacterium utilization of a novel plant-produced resource. Nat Biotech *15*, 363-368.

Scharf, B.E., Hynes, M.F., and Alexandre, G.M. (2016). Chemotaxis signaling systems in model beneficial plant-bacteria associations. Plant Mol Biol *90*, 549-559.

Schlüter, A., Szczepanowski, R., Pühler, A., and Top, E.M. (2007). Genomics of IncP-1 antibiotic resistance plasmids isolated from wastewater treatment plants provides evidence for a widely accessible drug resistance gene pool. FEMS Microbiol Rev *31*, 449-477.

Schmeisser, C., Liesegang, H., Krysciak, D., Bakkou, N., Le Quere, A., Wollherr, A., Heinemeyer, I., Morgenstern, B., Pommerening-Roser, A., Flores, M., *et al.* (2009). *Rhizobium* sp strain NGR234 possesses a remarkable number of secretion systems. Appl Environ Microbiol *75*, 4035-4045.

Schmid, J., Sieber, V., and Rehm, B. (2015). Bacterial exopolysaccharides: biosynthesis pathways and engineering strategies. Front Microbiol *6*, 496.

Schneiker, S., Keller, M., Dröge, M., Lanka, E., Pühler, A., and Selbitschka, W. (2001). The genetic organization and evolution of the broad host range mercury resistance plasmid pSB102 isolated from a microbial population residing in the rhizosphere of alfalfa. Nucleic Acids Res *29*, 5169-5181.

Slater, F.R., Bruce, K.D., Ellis, R.J., Lilley, A.K., and Turner, S.L. (2008). Heterogeneous selection in a spatially structured environment affects fitness tradeoffs of plasmid carriage in *Pseudomonads*. Appl Environ Microbiol *74*, 3189-3197.

Smalla, K., Haines, A.S., Jones, K., Krögerrecklenfort, E., Heuer, H., Schloter, M., and Thomas, C.M. (2006). Increased abundance of IncP-1 beta plasmids and mercury resistance genes in mercury-polluted river sediments: First discovery of IncP-1 beta plasmids with a complex mer transposon as the sole accessory element. Appl Environ Microbiol *72*, 7253-7259.

Smalla, K., Jechalke, S., and Top, E.M. (2015). Plasmid detection, characterization and ecology. Microbiol Spectr *3*, 10.1128/microbiolspec.PLAS-0038-2014.

Smillie, C., Garcillan-Barcia, M.P., Francia, M.V., Rocha, E.P.C., and de la Cruz, F. (2010). Mobility of plasmids. Microbiol Mol Biol Rev *74*, 434-452.

Spaepen, S., and Vanderleyden, J. (2011). Auxin and plant-microbe interactions. Cold Spring Harb Perspect Biol *3*, a001438.

Spaepen, S., Vanderleyden, J., and Remans, R. (2007). Indole-3-acetic acid in microbial and microorganism-plant signaling. FEMS Microbiol Rev *31*, 425-448.

Spencer, S.J., Tamminen, M.V., Preheim, S.P., Guo, M.T., Briggs, A.W., Brito, I.L., A Weitz, D., Pitkänen, L.K., Vigneault, F., Virta, M.P., *et al.* (2015). Massively parallel sequencing of single cells by epicPCR links functional genes with phylogenetic markers. The ISME Journal *10*, 427.

Sundin, G.W., Mayfield, C.T., Zhao, Y., Gunasekera, T.S., Foster, G.L., and Ullrich, M.S. (2004). Complete nucleotide sequence and analysis of pPSR1 (72,601 bp), a pPT23A-family plasmid from *Pseudomonas syringae* pv. *syringae* A2. Mol Genet Genomics *270*, 462-475.

Suzuki, K., Tanaka, K., Yamamoto, S., Kiyokawa, K., Moriguchi, K., and Yoshida, K. (2009). Ti and Ri Plasmids. In Microbial Megaplasmids, E. Schwartz, ed. (Springer Berlin Heidelberg), pp. 133-147.

Szegedi, E., Otten, L., and Czakó, M. (1992). Diverse types of tartrate plasmids in *Agrobacterium tumefaciens* biotype III strains. Mol Plant-Microbe Interact *5*, 435-438.

Szegedi, E., Süle, S., and Burr, T.J. (1999). *Agrobacterium vitis* strain F2/5 contains tartrate and octopine utilization plasmids which do not encode functions for tumor inhibition on grapevine. Journal of Phytopathology *147*, 554-558.

Taghavi, S., Barac, T., Greenberg, B., Borremans, B., Vangronsveld, J., and van der Lelie, D. (2005). Horizontal gene transfer to endogenous endophytic bacteria from poplar improves phytoremediation of toluene. Appl Environ Microbiol *71*, 8500-8505.

Tauch, A., Schneiker, S., Selbitschka, W., Pühler, A., van Overbeek, L.S., Smalla, K., Thomas, C.M., Bailey, M.J., Forney, L.J., Weightman, A., *et al.* (2002). The complete nucleotide sequence and environmental distribution of the cryptic, conjugative, broad-host-range plasmid pIPO2 isolated from bacteria of the wheat rhizosphere. Microbiology *148*, 1637-1653.

Tepfer, D., Goldmann, A., Pamboukdjian, N., Maille, M., Lepingle, A., Chevalier, D., Denarie, J., and Rosenberg, C. (1988). A plasmid of *Rhizobium meliloti* 41 encodes catabolism of two compounds from root exudate of *Calystegium sepium*. J Bacteriol *170*, 1153 - 1161.

Thomashow, M.F., Panagopoulos, C.G., Gordon, M.P., and Nester, E.W. (1980). Host range of *Agrobacterium tumefaciens* is determined by the Ti plasmid. Nature *283*, 794-796.

Top, E.M., Holben, W.E., and Forney, L.J. (1995). Characterization of diverse 2,4-dichlorophenoxyacetic acid-degradative plasmids isolated from soil by complementation. Appl Environ Microbiol *61*, 1691 - 1698.

Toussaint, A., and Merlin, C. (2002). Mobile elements as a combination of functional modules. Plasmid *47*, 26-35.

van der Auwera, G.A., Król, J.E., Suzuki, H., Foster, B., Van Houdt, R., Brown, C.J., Mergeay, M., and Top, E.M. (2009). Plasmids captured in *C. metallidurans* CH34: defining the PromA family of broad-host-range plasmids. Antonie Van Leeuwenhoek *96*, 193-204.

van Elsas, J.D., Mc Spadden Gardener, B.B., Wolters, A.C., and Smit, E. (1998). Isolation, characterization, and transfer of cryptic gene-mobilizing plasmids in the wheat rhizosphere. Appl Environ Microbiol *64*, 880 - 889.

van Elsas, J.D., Turner, S., and Bailey, M.J. (2003). Horizontal gene transfer in the phytosphere. New Phytol *157*, 525-537.

Vanbleu, E., Marchal, K., Lambrecht, M., Mathys, J., and Vanderleyden, J. (2004). Annotation of the pRhico plasmid of *Azospirillum brasilense* reveals its role in determining the outer surface composition. FEMS Microbiol Lett *232*, 165-172.

Velázquez, E., Peix, A., Zurdo-Pineiro, J.L., Palomo, J.L., Mateos, P.F., Rivas, R., Muñoz-Adelantado, E., Toro, N., Garcia-Benavides, P., and Martinez-Molina, E. (2005). The coexistence of symbiosis and pathogenicity-determining genes in *Rhizobium rhizogenes* strains enables them to induce nodules and tumors or hairy roots in plants. Mol Plant-Microbe Interact *18*, 1325-1332.

Veldman, K., Kant, A., Dierikx, C., van Essen-Zandbergen, A., Wit, B., and Mevius, D. (2014). Enterobacteriaceae resistant to third-generation cephalosporins and quinolones in fresh culinary herbs imported from Southeast Asia. Int J Food Microbiol *177*, 72-77.

Venturi, V., and Keel, C. (2016). Signaling in the rhizosphere. Trends Plant Sci *21*, 187-198.

Verraes, C., Van Boxstael, S., Van Meervenne, E., Van Coillie, E., Butaye, P., Catry, B., de Schaetzen, M.-A., Van Huffel, X., Imberechts, H., Dierick, K., *et al.* (2013). Antimicrobial resistance in the food chain: A review. Int J Env Res Public Health *10*, 2643-2669.

Viegas, C.A., Lilley, A.K., Bruce, K., and Bailey, M.J. (1997). Description of a novel plasmid replicative origin from a genetically distinct family of conjugative plasmids associated with phytosphere microflora. FEMS Microbiol Lett *149*, 121-127.

Villa, L., García-Fernández, A., Fortini, D., and Carattoli, A. (2010). Replicon sequence typing of IncF plasmids carrying virulence and resistance determinants. J Antimicrob Chemother *65*, 2518-2529.

Wabiko, H., Kagaya, M., and Sano, H. (1990). Various nopaline catabolism genes located outside the Ti-plasmids in *Agrobacterium tumefaciens* strains. Microbiology *136*, 97-103.

Wang, Y., Xiao, M., Geng, X., Liu, J., and Chen, J. (2007). Horizontal transfer of genetic determinants for degradation of phenol between the bacteria living in plant and its rhizosphere. Appl Microbiol Biotechnol *77*, 733-739.

Weingart, H., Völksch, B., and Ullrich, M.S. (1999). Comparison of ethylene production by *Pseudomonas syringae* and *Ralstonia solanacearum*. Phytopathology *89*, 360 - 365.

Wetzel, M.E., Kim, K.S., Miller, M., Olsen, G.J., and Farrand, S.K. (2014). Quorum-dependent mannopine-inducible conjugative transfer of an *Agrobacterium* opine-catabolic plasmid. J Bacteriol *196*, 1031-1044.

White, C.E., and Winans, S.C. (2007). Cell-cell communication in the plant pathogen *Agrobacterium tumefaciens*. Philosophical Transactions of the Royal Society B: Biological Sciences *362*, 1135-1148.

Wolters, B., Widyasari-Mehta, A., Kreuzig, R., and Smalla, K. (2016). Contaminations of organic fertilizers with antibiotic residues, resistance genes, and mobile genetic elements mirroring antibiotic use in livestock? Appl Microbiol Biotechnol *100*, 9343-9353.

Woodward, F.I., and Lomas, M.R. (2004). Vegetation dynamics - simulating responses to climatic change. Biol Rev Camb Philos Soc *79*, 643-670.

Zankari, E., Hasman, H., Cosentino, S., Vestergaard, M., Rasmussen, S., Lund, O., Aarestrup, F.M., and Larsen, M.V. (2012). Identification of acquired antimicrobial resistance genes. J Antimicrob Chemother *67*.

Zhang, M., Pereira e Silva, M.d.C., De Mares Maryam, C., and van Elsas, J.D. (2014). The mycosphere constitutes an arena for horizontal gene transfer with strong evolutionary implications for bacterial-fungal interactions. FEMS Microbiol Ecol *89*, 516-526.

Zhao, L., Fan, M., Zhang, D., Yang, R., Zhang, F., Xu, L., Wei, X., Shen, Y., and Wei, G. (2014). Distribution and diversity of rhizobia associated with wild soybean (*Glycine soja* Sieb. and Zucc.) in Northwest China. Syst Appl Microbiol *37*, 449-456.

Zhou, F., and Xu, Y. (2010). cBar: A computer program to distinguish plasmid-derived from chromosome-derived sequence fragments in metagenomics data. Bioinformatics *26*, 2051-2052.

Zhu, J., Oger, P.M., Schrammeijer, B., Hooykaas, P.J., Farrand, S.K., and Winans, S.C. (2000). The bases of crown gall tumorigenesis. J Bacteriol *182*, 3885-3895.

Zupan, J., Muth, T.R., Draper, O., and Zambryski, P. (2000). The transfer of DNA from *Agrobacterium tumefaciens* into plants: a feast of fundamental insights. The Plant Journal *23*, 11-28.

CHAPTER 3

Plant Immunity: The MTI-ETI Model and Beyond

Hanna Alhoraibi[1†], Jean Bigeard[2-3†], Naganand Rayapuram[1], Jean Colcombet[2-3] and Heribert Hirt[1*]

[1]Center for Desert Agriculture, Biological and Environmental Science and Engineering Division, 4700 King Abdullah University of Science and Technology, Thuwal, 23955-6900, Kingdom of Saudi Arabia
[2]Institute of Plant Sciences Paris-Saclay IPS2, CNRS, INRA, Université Paris-Sud, Université Evry, Université Paris-Saclay, Bâtiment 630, 91405 Orsay, France
[3]Institute of Plant Sciences Paris-Saclay IPS2, Paris Diderot, Sorbonne Paris-Cité, Bâtiment 630, 91405 Orsay, France
[†]These authors contributed equally to this manuscript

*heribert.hirt@kaust.edu.sa

DOI: https://doi.org/10.21775/9781912530007.03

Abstract
In plant-microbe interactions, a pathogenic microbe initially has to overcome preformed and subsequently induced plant defenses. One of the initial host-induced defense responses is microbe-associated molecular pattern (MAMP)-triggered immunity (MTI). Successful pathogens attenuate MTI by delivering various effectors that result in effector-triggered susceptibility and disease. However, some host plants developed mechanisms to detect effectors and can trigger effector-triggered immunity (ETI), thereby abrogating pathogen infection and propagation. Despite the wide acceptance of the above concepts, more and more accumulating evidence suggests that the distinction between MAMPs and effectors and MTI and ETI is often not given. This review discusses the complexity of MTI and ETI signaling networks and elaborates the current state of the art of defining MAMPs versus effectors and MTI versus ETI, but also discusses new findings that challenge the current dichotomy of these concepts.

Introduction

Plants are constantly exposed to a wide variety of adverse environmental conditions that can be broadly classified as biotic (bacteria, viruses, fungi, parasites, etc.) or abiotic stresses (drought, extreme temperature, chemicals, salinity, etc.). Attacks by pathogenic organisms constitute one of the most challenging situations during the life of a plant. Unlike animals, plants do not possess specialized mobile immune cells, but have nonetheless developed a rapid and effective immune system to survive and resist various pathogens. In addition, plants make use of preformed physical barriers, namely the cuticle and the cell wall, and constitutively produce antimicrobial compounds. The cuticle is a hydrophobic layer present on the external surface of the aerial epidermis of all land plants and is mainly composed of cutin and waxes (Yeats and Rose, 2013). Not only does it play a role in defense but it also acts as a barrier to transpirational water loss and as a protection against UV radiation. Although the cuticle is a good barrier against a number of pathogens, many fungal pathogens can penetrate the cuticle by mechanical rupture and secretion of cutinases that hydrolyze the cutin polyester (Longhi and Cambillau, 1999; Mendgen et al., 1996). In addition to the cuticle, the plant cell wall, which mainly consists of high molecular weight polysaccharides such as cellulose, hemicelluloses and pectin, glycosylated proteins and in certain cases lignin (Somerville et al., 2004), also protects plants against biotic aggressors. While fungal pathogens are equipped with cuticle and cell wall degrading enzymes to penetrate the epidermis, bacterial pathogens on the other hand do not typically enter plant tissues by directly penetrating the cuticle and cell wall. As a result they evolved strategies to enter the plant through a number of natural surface openings, such as stomata and through surface wounds caused by various environmental factors (Melotto et al., 2008).

Many plants produce two types of antimicrobial compounds, (i) preformed compounds also termed phytoanticipins that become toxic upon pathogen perception and (ii) induced compounds, such as camalexin produced following a pathogen attack (Arbona and Gomez-Cadenas, 2015; Osbourn, 1996). The induced compounds also include various proteins and small metabolites, such as phenolics, unsaturated lactones, saponins, cyanogenic glycosides and glucosinolates, that inhibit pathogen growth (Osbourn, 1996).

The two strategies, a preformed defense system and an inducible defense system, allow plants to withstand against a majority of plant pathogens, a phenomenon that is called non-host resistance. The inducible plant defense system has two layers, called microbe-associated molecular pattern (MAMP)-triggered immunity (MTI) and effector-triggered immunity (ETI). This review first discusses the signaling mechanisms occurring during MTI and ETI, and then discusses the current MTI-ETI dichotomy.

Signaling in MTI

The complex network of signaling events that occur during MTI has been exhaustively reviewed recently, with a specific emphasis on mitogen-activated protein kinases (MAPKs) (Bigeard et al., 2015). In the present review, we thus only briefly recapitulate the signaling in MTI.

MAMPs and PRRs

In MTI, the defense system is triggered by the detection and recognition of MAMPs, which are synthesized by pathogens and non-pathogens. Plants are also able to detect damage-associated molecular patterns (DAMPs) which are plant degradation products resulting from the action of invading pathogens, or endogenous peptides, constitutively present or newly synthesized, that are released by plants following a pathogen attack (Boller and Felix, 2009). Recognition of DAMPs also triggers responses similar to MTI responses. The most well characterized MAMP is flg22, a 22-amino-acid long epitope in the N-terminus of bacterial flagellin that is evolutionarily conserved and induces different defense responses (Zipfel, 2009; Zipfel et al., 2004). Other well-known examples of MAMPs that activate similar cellular responses are elf18 or elf26 (a conserved N-terminal portion of the bacterial elongation factor Tu), peptidoglycans (a component of bacterial cell walls), and chitin (a component of fungal cell walls) (Boller and Felix, 2009; Chisholm et al., 2006; Gomez-Gomez and Boller, 2002; Zipfel et al., 2006). MAMPs are recognized by Pattern-Recognition Receptors (PRRs) which are usually plasma membrane receptor-like kinases (RLKs) or receptor-like proteins (RLPs) with extracellular domains (Bohm et al., 2014a; Macho and Zipfel, 2014; Schwessinger and Ronald, 2012; Segonzac and Zipfel, 2011). Some of the best-known examples of PRRs are FLS2 (flagellin-sensitive 2), a leucine-rich receptor kinase, which recognizes flg22 (Gomez-Gomez and Boller, 2000) EF-Tu receptor (EFR) that perceives EF-Tu with the help of its minimal 18 amino acid epitope elf18 (Zipfel et al., 2006) and PEPR1, the receptor of the DAMP AtPep1 (Yamaguchi et al., 2006).

For MAMP perception and signal transduction, a number of PRRs have to associate with co-receptor RLKs (Monaghan and Zipfel, 2012), as shown for BAK1 (BRI1 associated receptor kinase 1), which can associate with a number of PRRs including FLS2, (Roux et al., 2011; Segonzac and Zipfel, 2011). FLS2 also associates with the RLK BIK1 (Botrytis-induced kinase 1) and related PBL (PBS1-like) proteins, which are rapidly released from FLS2 upon flg22 binding (Liu et al., 2013; Lu et al., 2010; Zhang et al., 2010). MAMP perception induces very rapid auto- and trans-phosphorylation reactions of these interacting proteins (Lu et al., 2010; Schulze et al., 2010), followed by a complex sequence of choreographed events (Figure 1).

Early events in MTI signaling

Among the earliest responses to MAMP/DAMP perception is an influx of extracellular Ca^{2+} ions into the cytosol (Jeworutzki et al., 2010; Nomura et al., 2012; Ranf et al., 2011), inducing the opening of other membrane channels (influx of H^+, efflux of K^+, Cl^- and nitrate) which lead to an extracellular alkalinization and a depolarization of the plasma membrane (Jeworutzki et al., 2010). In addition to Ca^{2+} ion fluxes, a very early event following MAMP/DAMP recognition is the production of reactive oxygen species (ROS) (Chinchilla et al., 2007); (Nuhse et al., 2000); (Ranf et al., 2011) mainly by the plasma membrane-localized NADPH oxidase RBOHD (respiratory burst oxidase homolog D) (Nuhse et al., 2007; Ranf et al., 2011). Upon MAMP perception, RBOHD is phosphorylated by Ca^{2+}-induced CDPKs (calcium-dependent protein kinases) and BIK1 on different residues, which are all required for activation of the NADPH oxidase (Boudsocq et al., 2010; Dubiella et al., 2013; Kadota et al., 2014). Ca^{2+} itself also regulates RBOHD through direct binding to the N-terminal EF-hand

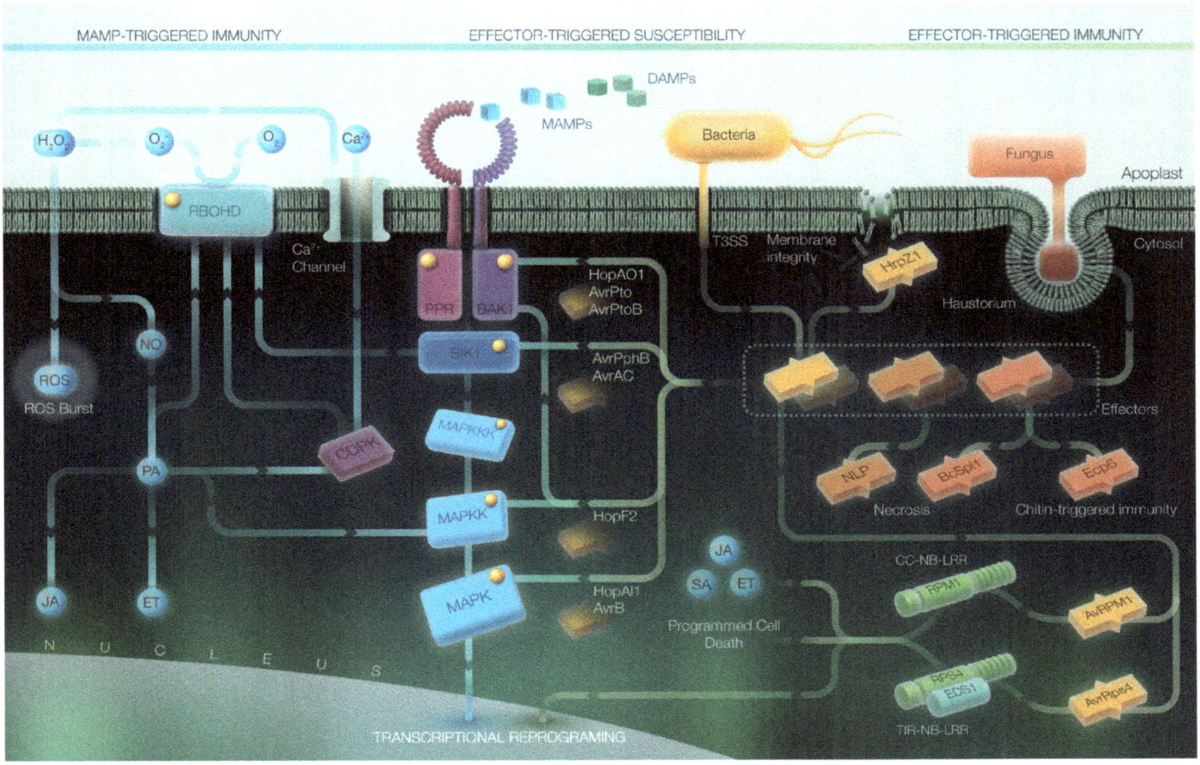

Figure 1. Inducible defense systems in plants. PRRs perceive the MAMPs and recruit BAK1 and BIK1 to induce MTI involving notably MAPK modules. Plants also detect DAMPs that are degradation products and trigger responses similar to the MTI responses. This is accompanied by a ROS burst via the NADPH oxidase RBOHD which in turn is phosphorylated by Ca^{2+}-induced CDPKs. RNS, such as NO, are required for generation of PA via both PLD and PLC/DGK pathways. PA can interact and modulate the activity of CDPKs, MAPKs and RBOHD/F (Zhang et al., 2009) and can regulate production of JA and ET. Bacterial and fungal pathogens may deliver effectors via the T3SS and haustoria, respectively, that block MTI. Plants evolved CNLs or TNLs to nullify the effect of the effectors leading to a stronger immune response termed ETI that involves transcriptional reprogramming, programmed cell death and increased levels of the hormones SA, JA and ET. The effector HopAO1 dephosphorylates the PRR and suppresses subsequent immune response. The effectors AvrPto and AvrPtoB suppress immunity by acting directly on the MAMP receptors via inhibiting BAK1 kinase activity while the effectors AvrPphB and AvrAC inhibit the response by cleaving or uridylylating BIK1. The effectors AvrB, HopAI1 and HopF2 directly target different components of the MAPK cascades. The TIR-NB-LRR, RPS4 recognizes the effector AvrRPS4 and redistributes the EDS1-RPS4 complex between the nucleus and cytoplasm to induce defense responses. An example of CC-NB-LRR is RPM1 that recognizes AvrRpm1. Some MAMPs, such as flg22, bacterial LPS and harpin (HrpZ1) act as effectors too. Similarly, certain effector proteins such as NLPs, BcSpl1 and LysM domain containing proteins such as Ecp6 have a more widespread occurrence and function as MAMPs too.

motifs of the protein (Ogasawara et al., 2008). RBOHD produces membrane-impermeable superoxide ($O_2^{\cdot -}$) in the apoplast, which is converted into hydrogen peroxide (H_2O_2) by superoxide dismutases. In contrast to other ROS, H_2O_2 is relatively stable and membrane-permeable and can enter the cytosol and different organelles of plant cells. However, NADPH oxidases are not the only source for ROS, but multiple ROS sources seem to be involved in a complex temporal and spatial coordination (Baxter et al., 2014; Gross et al., 2013).

Other than Ca^{2+} and ROS, reactive nitrogen species (RNS), such as nitric oxide, were shown to be involved at different steps of MAMP/DAMP signaling, e.g. via inhibition of RBOHD or regulation of NPR1 (non-expresser of PR genes 1), a master regulator of defense gene expression, which both become nitrosylated on cysteine residues (Tada et al., 2008; Yun et al., 2011). Some lipid derivatives, such as phosphatidic acid (PA) and ceramides, were also proposed to function as signaling molecules upon pathogen infection (Okazaki and Saito, 2014). MAMP/DAMP-induced NO production is partly also required for PA generation via both the phospholipase D (PLD) and phospholipase C/ diacylglycerol kinase (PLC/DGK) pathways (Raho et al., 2011). PA can interact and modulate the activity of CDPKs (Farmer and Choi, 1999; Szczegielniak et al., 2005) MAPKs (Testerink et al., 2007), RBOHD/F (Zhang et al., 2009) and can regulate production of jasmonic acid (JA) and ethylene (ET) (Nakano et al., 2013; Testerink et al., 2008; Testerink et al., 2007; Wang et al., 2000).

Activation of protein kinases

Besides the very rapid auto- and trans-phosphorylation reactions at the level of the receptor complexes, other protein kinases get activated in a matter of minutes and most of these belong to the CDPK and MAPK protein kinases and are key elements in regulating defense at the level of the transcriptional and metabolic responses (Boudsocq et al., 2010; Frei dit Frey et al., 2014; Lassowskat et al., 2014).

Among the CDPKs, CPK4, 5, 6 and 11 are rapidly activated upon flg22 signaling (Boudsocq et al., 2010) and were shown to regulate ROS production via phosphorylation of NADPH oxidase RBOHD, transcriptional reprogramming and resistance to the bacterial pathogen *Pseudomonas syringae* pv tomato DC3000 (*Pst* DC3000) (Boudsocq et al., 2010; Romeis and Herde, 2014). CDPK substrates include the important regulators RBOHD and ACS2 (Boudsocq and Sheen, 2013; Dubiella et al., 2013; Kamiyoshihara et al., 2010; Schulz et al., 2013). Another important CDPK seems to be CPK28, as loss of function *cpk28* mutant accumulates high levels of the plasma membrane associated cytoplasmic kinase BIK1 and exhibits strong MAMP-triggered responses (Monaghan et al., 2014). CPK28 is genetically upstream of the MAMP-triggered Ca^{2+} burst and negatively regulates BIK1 by phosphorylation that marks it for ubiquitination and subsequent degradation (Monaghan et al., 2015).

MAMP/DAMP perception activates a number of MAPKs, including the following members of the gene family of 20 MAPKs MPK1, 3, 4, 6, 11 and 13 (Bethke et al., 2012; Nitta et al., 2014; Nuhse et al., 2000; Zipfel et al., 2006). MAPK kinase kinases (MAPKKKs), MAPK kinases (MAPKKs) and MAPKs constitute functional signaling modules. Two signaling modules have been defined to date upon MAMP perception, namely MKK4/MKK5-MPK3/ MPK6 (Asai et al., 2002; Ren et al., 2002) and MEKK1-MKK1/MKK2-MPK4 (Berriri et al., 2012; Gao et al., 2008; Hadiarto et al., 2006; Huang et al., 2000; Ichimura et al., 2006; Ichimura et al., 1998; Matsuoka et al., 2002; Mizoguchi et al., 1998; Nakagami et al., 2006; Petersen et al., 2000; Qiu et al., 2008; Suarez-Rodriguez et al., 2007; Teige et al., 2004). In the current model, both modules positively regulate defense responses (Berriri et al., 2012; Kong et al., 2012; Pitzschke et al., 2009; Rasmussen et al., 2012; Su et al., 2013; Zhang et al., 2012; Zhao et al., 2014). However, the molecular link between the PRRs and these MAPK pathways remains to be elucidated. Regarding MPK1, MPK11 and MPK13, their upstream MAPKKs and MAPKKKs have not been identified yet. Besides, *mpk1*, *mpk11* and *mpk13* mutants do not show altered resistance to a bacterial pathogen suggesting a functional redundancy among the MAPKs and does not preclude yet the identification of their roles in plant immunity (Nitta et al., 2014). MPK3, MPK4 and MPK6 phosphorylate specific and redundant substrates to control many cellular responses. For example, the ET-related ERF104 is specifically targeted by MPK6 (Bethke et al., 2009), while ACS2 and ACS6 are phosphorylated by both MPK3 and MPK6 (Han et al., 2010; Liu and Zhang, 2004). The number of identified substrates is constantly growing, highlighting the importance of MAPKs during MTI. Additionally, MAPKs also play important roles in abiotic stresses and development (Colcombet and Hirt, 2008; Rodriguez et al., 2010).

Role of hormones in MTI signaling

In response to infection by biotrophic and hemibiotrophic pathogens, salicylic acid (SA) plays a pivotal role in plant defense by regulating its downstream components. Elevated levels of SA cause nuclear accumulation of NPR1 (SA receptor), which is subsequently degraded to mediate systemic acquired resistance (SAR) (Vlot et al., 2009; Wu et al., 2012). SA is also associated with the accumulation of antimicrobial pathogenesis-related (PR) proteins (Moore et al., 2011). MPK3 and to a lesser extent MPK6 have been proposed to play an important role in SA-mediated priming and enhancing defense gene activation and resistance (Beckers et al., 2009). On the other hand, the MPK4 cascade negatively regulates SA signaling and mutants of this cascade exhibit SA accumulation, constitutive pathogenesis-related gene expression and SAR (Petersen et al., 2000). In the case of necrotrophic pathogen infections, JA and ET are induced. The two tobacco orthologs of MAPKs, WIPK and SIPK, regulate the levels of JA in wounded tobacco plants (Seo et al., 2007). Both MAPKs are required but not sufficient to induce JA production (Kim et al., 2003). MKK3 and MPK6 negatively regulate *AtMYC2* in both JA-dependent gene expression and inhibition of root growth, which indicates a possible role for the MKK3-MPK6 cascade in JA signal transduction (Takahashi et al., 2007). Moreover, it has been found that *mpk4* mutant plants are defective in inducing JA and ET defense marker genes such as *PDF1.2* in response to JA. MPK4 positively regulates JA/ET-inducible gene responses through the defense regulators EDS1 and PAD4 independently of its negative regulation of SA biosynthesis (Kong et al., 2012; Petersen et al., 2000). Thus, MPK4 is proposed to be required for the balance between SA and JA/ET related defense (Brodersen et al., 2006).

Reprogramming of gene expression

MAMPs/DAMPs trigger a massive and dynamic reprogramming of plant genome expression. Several thousand genes are affected by flg22 perception (Denoux et al., 2008). Transcriptional reprogramming of defense hormone signaling as well as the synthesis of antimicrobial compounds becomes apparent after 1 hour (Tsuda et al., 2009). Later, genes mainly involved in SA-mediated secretory processes and senescence are prominently affected (Denoux et al., 2008). The chloroplast resident calcium-sensing receptor (CAS) acts upstream of SA accumulation and is involved in MAMP-induced expression of defense genes while also suppressing chloroplast gene expression thus allowing chloroplast mediated transcriptional reprogramming in cytoplasmic-nuclear plant immune responses (Nomura et al., 2012). Numerous transcription factors are thus involved in plant immunity (Alves et al., 2013; Ambawat et al., 2013; An and Mou, 2013; Eulgem and Somssich, 2007; Gatz, 2013; Gutterson and Reuber, 2004; Nuruzzaman et al., 2013; Pandey and Somssich, 2009; Puranik et al., 2012). In addition, it is becoming more and more clear that chromatin remodelers and modifiers also contribute strongly to transcriptional regulation of defense (Berr et al., 2012; Dowen et al., 2012; Ma et al., 2011; Yu et al., 2013).

Signaling in ETI

Effectors and R proteins

Effectors are molecules produced by plant pathogens and function as virulence factors to mediate infection of specific plant species or varieties. These molecules can be proteins, nucleic acids, carbohydrates or metabolites. Effectors can have different effects such as inhibiting MTI or ETI and can be secreted either into the extracellular matrix or directly delivered into the plant cell (Hogenhout et al., 2009). Many effectors are proteins that are injected into plant host cells through bacterial type III secretion systems (T3SS) (Feng and Zhou, 2012; Hann et al., 2010; Lohou et al., 2013). The genomes of plant pathogens may contain a considerable number of effectors, as evidenced from analysis of the model bacterial pathogen *P. syringae* that contains between 30 - 50 genes coding for effector proteins (Buell et al., 2003). A significant number of effectors target components of PRR immune complexes or the downstream signaling cascades (Feng and Zhou, 2012; Mukhtar et al., 2011). For instance, the *P. syringae* effector HopAI1, a phosphothreonine lyase, directly targets and inactivates MPK3, MPK4 and MPK6 by dephosphorylating these kinases (Zhang et al., 2007). The HopF2 effector inactivates MKK5 and probably other MKKs to inhibit MAPK signaling to suppress downstream defense responses (Wang et al., 2010). Another effector, HopAO1, a protein tyrosine phosphatase targets the phosphorylation on a specific tyrosine residue on the PRR EFR (Y836) (and also probably FLS2) to inhibit ligand-induced activation of the PRR and suppresses the subsequent immune response (Espinosa et al., 2003) (Macho et al., 2014). The effector AvrB was also reported to regulate hormone signaling by inducing MPK4 phosphorylation thus enhancing plant susceptibility (Cui et al., 2010). An alternative way to suppress immunity is by targeting components upstream of MAPKs by pathogen effectors such as the MAMP receptors FLS2, EFR, and CERK1 by AvrPto and AvrPtoB, by inhibiting BAK1 kinase activity via interaction with AvrPtoB, or by cleaving and uridylylating BIK1 by AvrPphB and AvrAC to inhibit MTI signaling (Meng and Zhang, 2013). The *Agrobacterium* T-DNA associated virulence protein VirE2 together with the host cell transcription factor VIP1 binds to the nuclear import machinery to transfer the T-DNA to the nucleus. For this process to occur, VIP1 needs to be phosphorylated by MPK3 to translocate from the cytoplasm to the nucleus. The bacterial VirF effector contains an F-box motif and targets VirE2 and VIP1 for proteosomal degradation (Djamei et al., 2007; Tzfira et al., 2004). Plant pathogens not only produce protein effectors, but also small molecules, such as the polyketide coronatine, which structurally and functionally mimics the active plant hormone conjugate JA-isoleucine (JA-Ile). Coronatine is secreted by several pathovars of *P. syringae* and contributes to virulence by antagonizing SA-mediated host responses (Weiler et al., 1994; Xin and He, 2013). However, yet another strategy is the production of small RNAs to hijack the plant RNA interference (RNAi) machinery, as recently shown for *Botrytis cinerea* via the host protein AGO1, which in turn silences host immunity genes (Weiberg et al., 2013).

Plant R proteins are intracellular receptors that detect the presence of pathogen effectors in the host cell. Most of them are nucleotide-binding domain and leucine-rich repeat (NB-LRR or NLR) proteins (Jacob et al., 2013; Maekawa et al., 2011; Qi and Innes, 2013). Briefly, NLR proteins are divided into two groups depending on their N-terminal structures: CNL (CC-NB-LRR) with an N-terminal coiled-coil domain and TNL (TIR-NB-LRR) with an N-terminal Toll/interleukin-1 receptor domain (TIR). Some NLRs contain domains termed integrated decoys that recognize effectors from pathogens. These were found in multiple plant families indicating their functional significance and conservation. Across plant lineages, domains already known to be implicated in pathogen defense such as in the case of RIN4, NPR1 and Zinc Finger BED type protein (ZBED) have been integrated into NLR proteins (Kroj et al., 2016; Sarris et al., 2016). They can have different sub-cellular localizations (plasma membrane-associated, cytosolic, nuclear, etc.) and intracellular shuttling is important for some NLR proteins to fulfill their functions (Dowen et al., 2009; Garcia et al., 2010; Wirthmueller et al., 2007). An accumulation of plant NLRs leads to autoimmunity and so NLR homeostasis is tightly regulated at multiple levels (Huang et al., 2016; Kadota et al., 2010; Shirasu, 2009; Takken and Goverse, 2012). R proteins detect pathogen effectors in three possible ways, either through direct physical interaction (Dodds and Rathjen, 2010) or by sensing effector-induced modification of other plant proteins termed as the guardee/decoy model (Dangl and Jones, 2001; van der Hoorn and Kamoun, 2008) or via a third method termed the integrated decoy model where in the R proteins have incorporated a decoy domain into their structure (Cesari et al., 2014). They are thought to be auto-inhibited and activated upon ligand binding. LRR domains mostly seem to be responsible for effector recognition, while the TIR or CC domains function in signal transduction (Qi and Innes, 2013) (Figure1).

Signaling mechanisms by some NLRs

The mechanism of effector recognition is now known for a number of NB-LRR-effector pairs. The *P. syringae* type III effector AvrRps4 is recognized by the TIR-NB-LRR receptor RPS4 (Gassmann et al., 1999). RPS4 distributes between endomembranes and nuclei both in healthy and AvrRps4-triggered tissues (Wirthmueller et al., 2007). Like all TIR-type NB-LRRs, RPS4 requires interaction with the basal defense regulator enhanced disease susceptibility 1 (EDS1), a lipase-like protein, to activate ETI, and a coordinated nucleo-cytoplasmic partitioning of EDS1-RPS4 complex is necessary to trigger the full set of immune responses (Bhattacharjee et al., 2011; Heidrich et al., 2011). In fact, forced nuclear localization of the AvrRps4 effector is sufficient to induce RPS4-mediated bacterial growth inhibition but hinders RPS4-mediated HR, while forced cytoplasmic localization of AvrRps4 decreases RPS4-mediated bacterial growth inhibition but only moderately reduces RPS4-mediated HR (Heidrich et al., 2011). These results suggest that a single NLR may activate distinct signaling pathways in the cytoplasm and nucleus and that cell death and the restriction of pathogen growth are two separate phenotypes. Recognition of the bacterial effector proteins AvrB and AvrRpm1 occurs via RPM1, a CC-NB-LRR (Grant et al., 1995). RPM1 is plasma membrane-localized in both the inactive and active forms (Boyes et al., 1998; Gao et al., 2011) and in this case nuclear re-localization is not required for RPM1-mediated defense responses or the induction of HR (Gao et al., 2011), suggesting that ETI signaling can function by different mechanisms.

RPS2 and RPM1 are two plasma membrane-associated CC-NB-LRRs. In their case, the signaling mechanisms are well documented. Using inhibitors, RPS2- and RPM1-mediated signaling was shown to depend on the sequential production of PA by PLC/DGK and the influx of extracellular Ca^{2+} followed by production of ROS and PA via PLD (Andersson et al., 2006). The influx of extracellular and release of internal Ca^{2+} then results in a complex system of CDPK activations (Gao et al., 2013). The immune response is orchestrated by defense gene expression via phosphorylation of the WRKY8/28/48 transcription factors by CPK4/5/6/11, the induction of ROS production through phosphorylation of the NADPH oxidases RBOHD and F by CPK1/2/4/11. ETI mediated by RPS2 and RPM1 was also shown to be reduced in a calcium-sensing receptor (CAS) mutant, as revealed by reduced ROS and NO production and a delayed and suppressed HR cell death and demonstrating the role of chloroplast signaling in these NLR-triggered responses (Nomura et al., 2012). The contribution of the SA, JA and ET hormone pathways in ETI was estimated by measuring the relative growth of *Pst* DC3000 strains expressing either of the effectors AvrRpt2, AvrRpm1 or AvrPphB, which are recognized by the CC-NB-LRRs RPS2, RPM1 and RPS5, respectively (Tsuda et al., 2009). While the absence of individual phytohormone signaling pathways had no dramatic effect on the ETI response, the defense responses decreased by up to 80% in the combined absence of the SA, JA and ET signaling pathways. These results demonstrate the overlapping contributions of SA, JA and ET signaling pathways in NLR-mediated immune responses but also the variability in phytohormone-dependency of different NLRs to trigger defenses. (Tao et al., 2003).

Besides the signaling mechanisms described above, other signaling routes are observed in the case of several TIR-NB-LRRs. Indeed, some TIR-NB-LRR-interacting proteins such as EDS1, suppressor of rps4-RLD1 (SRFR1) and Topless-related 1 (TPR1) probably represent signaling complexes that act as transcriptional regulators (Bhattacharjee et al., 2013; Bhattacharjee et al., 2011; Kim et al., 2014b; Zhu et al., 2010). Interestingly, chromatin regulation also seems to contribute to transcription regulation in ETI as seen for example by the histone deacetylase 19 (HDA19) that forms a complex with TPR1 (Ma et al., 2011; Zhu et al., 2010). Some CC-NB-LRRs also interact with transcription factors, such as the activated barley (*Hordeum vulgare*) MLA10 which induces MYB6-dependent gene regulation and the rice Pb1 which interacts with WRKY45 to prevent its ubiquitin-proteasome degradation (Bhattacharjee et al., 2013; Chang et al., 2013; Inoue et al., 2013). It thus seems that direct R gene-mediated transcriptional regulation might in some cases also be at the heart of ETI.

Signaling mechanisms by RPW8.2

Resistance to powdery mildew (*Golovinomyces orontii*) requires the atypical R gene RPW8.2 (resistance to powdery mildew 8.2). Although RPW8.2 shows no similarity to other NLRs, RPW8.2 also requires EDS1 to induce HR and as well as SA, PAD4, EDS5 and NPR1 (Xiao et al., 2005). Upon infection by *Golovinomyces orontii*, the transcription of *RPW8.2* is strongly induced and RPW8.2 protein is carried on VAMP721/722 vesicles to the extrahaustorial membrane (EHM) independently of SA signaling (Kim et al., 2014a; Wang et al., 2009). RPW8.2 activates an EDS1 and SA signaling-dependent defense process that concomitantly enhances callose deposition and accumulation of H_2O_2 at the haustorial interface (Wang et al., 2009). In addition, RPW8.2 interacts with the 14-3-3 isoform lambda protein which may positively regulate RPW8.2 (Yang et al., 2009). 14-3-3 proteins were also shown to be involved in ETI in other systems (Oh and Martin, 2011; Oh et al., 2010; Teper et al., 2014).

Observations going beyond the MTI-ETI model

The strict separation of MTI and ETI results in the assumption that MAMPs are very conserved molecules that are widely detected while effectors are variable and only sensed by specific hosts. However, accumulating evidences suggest that the story of MAMPs and effectors is more complicated. The disappearing boundaries differentiating MTI-ETI and the concept of invasion model of plant immune system were put forth by Thomma and co-workers (Cook et al., 2015; Thomma et al., 2011).

Effectors and R proteins with broader scopes

Recently, it has become apparent that many effector proteins have a more widespread occurrence, which would equally qualify them as MAMPs. A good example is the necrosis and ET-inducing peptide 1 (Nep1) that was originally identified from *Fusarium oxysporum* (Bailey, 1995). Moreover, various Nep1-like proteins (NLPs) are encoded by bacteria, fungi and oomycetes and positively contribute to virulence of these pathogens (Gijzen and Nurnberger, 2006; Ottmann et al., 2009). Interestingly, a conserved amino acid motif was recently identified in NLPs that serves as a potent MAMP (Bohm et al., 2014b; Oome et al., 2014), thereby NLPs fulfill all the criteria for being effectors and MAMPs. Another example is BcSpl1, an effector protein required for full virulence of the necrotrophic fungus *Botrytis cinerea* (Frias et al., 2014; Frias et al., 2011). Two conserved peptide stretches of BcSpl1 can induce host defense and cell death. Since the two conserved regions are present in all BcSpl1 family members and belong to a highly conserved protein effector family in fungi, BcSpl1 can be classified as an effector and also as a MAMP (Frias et al., 2014). The *Pseudomonas syringae pv. phaseolicola* protein HrpZ1 has the ability to form ion-conducting pores and these pores have been proposed to facilitate delivery of effectors into the plant, thus functioning as a virulence factor that affects host membrane integrity. HrpZ1, especially the C-terminal fragment, is a MAMP that triggers MTI-like responses in a variety of plants, thus exhibiting a dual role in plant immunity during infection (Engelhardt et al., 2009). Yet another example of a fungal effector that also behaves as a MAMP is the well characterized LysM effector Ecp6 (extra cellular protein 6). Ecp6 interferes with chitin-triggered activation of host immune responses by sequestering chitin fragments thereby qualifying as an effector. Interestingly, Ecp6 also competes with the plant LysM domain-containing chitin receptor CEBiP for binding chitin fragments. Ecp6 is found in all strains of *Cladosporium fulvum* with very little sequence variation. The widespread occurrence and functional conservation of LysM effectors is reminiscent of MAMPs and qualifies them to be designated as MAMPs (de Jonge and Thomma, 2009; de Jonge et al., 2010; Thomma et al., 2011).

Along the same line, R genes have been mostly thought of being receptors with specificity to a particular pathosystem. However, the NLR Rxo1 of maize not only confers resistance to *Burkholderia andropogonis*, the causal agent of maize stripe disease, but also to the unrelated bacterial rice pathogen *Xanthomonas oryzae* pv. oryzicola, which triggers ETI upon recognition of the type III effector protein AvrRxo (Zhao et al., 2004). Similarly, the physically linked NLR pair RRS1 and RPS4 confers resistance to Brassicaceae to the fungal pathogen *Colletotrichum higginsianum*, the broad-host range bacterial wilt pathogen *Ralstonia solanacearum*, and the bacterial pathogen *P. syringae* (Narusaka et al., 2009). In another example, the NLR immune receptor of tomato Mi-1.2 confers resistance to phloem-feeding insects as well as root-knot nematodes (Rossi et al., 1998; Vos et al., 1998). In a screen of 171 predicted bacterial effectors from *Pseudomonas, Ralstonia*, and *Xanthomonas* expressed in 59 plants from four plant families, it was found that each plant responded to an average of 19 effectors. Interestingly, the necrotic response to an effector was generally not taxonomically defined (Wroblewski et al., 2009). Taken together, these examples demonstrate that resistance conferred by NLR immune receptors is not necessarily restricted to a single pathosystem. Although some NLRs may directly perceive effectors, broadly detected effectors are likely perceived indirectly because they induce DAMPs or modify host targets that are guarded by R proteins (the guard model) (Van der Biezen and Jones, 1998). Nevertheless, broad detection of effectors by NLRs is conceptually similar to MAMP recognition by PRRs.

MAMPs and PRRs with reduced scopes

Conversely for MAMPs and PRRs, purified flagella and flg22 can induce immune responses in many different plant species, but with different efficiencies (Felix et al., 1999). The naturally occurring variation in the flagellin amino acid sequences of the bacterial pathogen *Xanthomonas campestris* pv. *campestris* (Xcc) correlates with its pathovar-dependent potential of defense response induction (Sun et al., 2006). Similarly, *R. solanacearum* strain K60 and *Pseudomonas cannabina* pv. *alisalensis* (*Pcal*) strain ES4326 show convincing correlations between their respective flg22 epitope sequence variations and the induced immune responses (Clarke et al., 2013; Pfund et al., 2004). Additional evidence comes from the analysis of the evolution of the flagellin sequences in natural populations of *P. syringae* pathovars. Here, the variation of the flg22 epitope sequences clearly indicated the evolution of the pathogenic potential to escape MAMP detection. Moreover, a second, 28-amino acid immunogenic region of flagellin, termed flgII-28, induced defense responses in tomato, and both the flg22 and flgII-28 peptides contribute to the ROS burst (Cai et al., 2011). Interestingly, the flgII-28 epitope induced immune responses in various solanaceous species but not in a variety of plants from five other families, suggesting that the perception system for the flgII-28 epitope is a rather recent specific achievement of solanaceae (Clarke et al., 2013). The recent characterization of the orthologous grape flagellin receptor VvFLS2 indicates that the flagellin encoded by a grape-adapted, plant growth-promoting rhizobacterium (PGPR), *Burkholderia phytofirmans*, elicits a weaker immune response on grape compared with flg22, which is specifically conditioned by the VvFLS2 receptor (Trda et al., 2014).

Correlatively, variation for flagellin perception is also conditioned by variation in the plant receptor FLS2 (Gomez-Gomez and Boller, 2000). The *Arabidopsis thaliana* accession Ws-0 does not respond to flg22 nor does it contain a functional FLS2 allele as it carries a point mutation that results in a stop codon in the kinase domain of FLS2 (Bauer et al., 2001; Zipfel et al., 2004), and genotypes in closely related *Arabidopsis lyrata*, *Cardamine hirsuta*, and additional Brassicaceae species do not bind the flg22 epitope (Vetter et al., 2012). The tomato and *Nicotiana benthamiana* orthologs of AtFLS2 display species-specific, receptor-dependent variation for flagellin perception (Robatzek et al., 2007).

Protease activated immune signaling

The bacterial pathogen *Pseudomonas aeruginosa* strain P14 secretes a PvdS-regulated lysyl class serine protease (protease IV) that elicits a strong immune response comparable to the response elicited by flg22 in terms of the activation of MPK3 and MPK6 but not MPK4, oxidative burst, expression of defense related genes and protects Arabidopsis plants from *Pst* DC3000 infection. The activation of MAPKs in response to protease IV requires the Gα, Gβ and Gγ subunits of the heterotrimeric G-protein complexes. The receptor activated C kinase 1 (RACK1) acts as a scaffold and connects G-protein signaling to the downstream MAPK cascade. This module, involving the protease-G-protein-RACK1-MAPK cascades, forms a novel protease mediated immune signaling pathway distinct from the ones previously described (Cheng et al., 2015).

Signaling similarities in MTI and ETI

While MTI employs a core set of signaling events, ETI does not seem to be mediated by such a core set of signaling components. Rather, and dependent on the activated R protein, different subsets of signaling elements are solicited which nonetheless finally result in an efficient immune response. In addition to sharing a number of similar events between MTI and ETI, such as the production of ROS via the NADPH oxidase RBOHD, a calcium burst, the synthesis of PA and NO, MTI and ETI also employ common signaling pathways, as exemplified by the MAPK and CDPK cascades.

Before R protein-mediated signaling can occur in ETI, first the R proteins need to detect the presence of the effector. Several reports indicate that effectors injected by *P. syringae* are detected in plant cells (*Arabidopsis*, tobacco, tomato) 2 to 3 h post-inoculation (Mudgett and Staskawicz, 1999; Schechter et al., 2004). These results are consistent with another pathosystem, *Xanthomonas campestris pv. vesicatoria*/pepper plants (Casper-Lindley et al., 2002). These results as well as the comparison of ROS burst in *Arabidopsis* leaves with live *Pseudomonas* and flg22 elicitation (Smith and Heese, 2014) indicate that ETI may be initiated rapidly during the infection process and that MTI and ETI probably occur very close in time.

However, in contrast to the transient MAPK activation in MTI, induction of the expression of effectors by estradiol-inducible promoters resulted in the activation of MPK3 and MPK6 for several hours (Tsuda et al., 2013). Interestingly, Qi et al. showed that FLS2 can form a complex with the R proteins RPS2, RPM1 and RPS5 which are all plasma membrane-localized CC-NB-LRRs (Qi et al., 2011). The biological significance of this MTI-ETI receptor R protein complex is however currently not known but the existence of such a complex due to the possibility of several shared components between MTI and ETI is not far fetched.

Signaling by the three phytohormones SA, JA and ET has shown to be activated in some cases of both MTI as well as ETI (Tsuda and Katagiri, 2010; Tsuda et al., 2009). A positive role for the three hormones in flg22-triggered immunity (MTI) and AvrRpt2-triggered immunity (ETI) was demonstrated using an *Arabidopsis* dde2/ein2/pad4/sid2 quadruple mutant, which is a loss of function mutant of essential components involved in JA, ET and SA signaling. This again reinforces the fact that MTI and ETI share common signaling networks but use them in specific circumstances.

The transcriptional reprogramming integrates a large part of the upstream signaling inputs mediated mainly by the protein kinases and allows the implementation of induced defense mechanisms. It is dynamically regulated and it involves numerous transcription factors and chromatin regulators (Moore et al., 2011). Importantly, the differentially expressed genes during MTI and ETI are identical, but differ in quantity and kinetics (Tao et al., 2003). The stronger response in ETI suggests that ETI employs some of the same signaling components as MTI but results in higher expression of its target genes. Considering the hypothesis of MTI having evolved before ETI, these observations imply that ETI acquired R proteins during the course of evolution while adopting several signaling components of the MTI pathway (Tsuda and Katagiri, 2010).

Conclusions

Plant defense against pathogens is based on both preformed and induced defenses. Preformed defenses hinder pathogen entry and are among the main contributors to non-host resistance. Induced defenses are activated after perception of invading pathogens via two classically separate routes, MTI and ETI. Pathogen perception that leads to MTI is mediated by recognition of MAMPs by PRRs, whereby ETI recognizes effectors by intracellular R proteins. Although the MTI-ETI model constitutes an important and useful concept, the separation between MAMPs and effectors, and between PRRs and R proteins is not always so clear, and thus the dichotomy of MTI-ETI cannot be maintained but is rather a continuum between MTI and ETI.

Acknowledgements

All authors contributed to this work. HA, JB, NR and HH wrote the article and JB, JC and HH revised the work. We would like to thank Carlos Xavier Pita from KAUST for help with scientific illustrations. We would like to apologize for not citing several research articles because of space limitations. The work by NR, HA and HH is supported from funds which are made available to HH by KAUST. The IPS2 benefits from the support of the LabEx Saclay Plant Sciences-SPS (ANR-10-LABX-0040-SPS).

References

Alves, M.S., Dadalto, S.P., Goncalves, A.B., De Souza, G.B., Barros, V.A., and Fietto, L.G. (2013). Plant bZIP transcription factors responsive to pathogens: a review. International journal of molecular sciences *14*, 7815-7828. doi: 10.3390/ijms14047815.

Ambawat, S., Sharma, P., Yadav, N.R., and Yadav, R.C. (2013). MYB transcription factor genes as regulators for plant responses: an overview. Physiology and molecular biology of plants : an international journal of functional plant biology *19*, 307-321. doi: 10.1007/s12298-013-0179-1.

An, C., and Mou, Z. (2013). The function of the Mediator complex in plant immunity. Plant signaling and behavior *8*, e23182. doi: 10.4161/psb.23182.

Andersson, M.X., Kourtchenko, O., Dangl, J.L., Mackey, D., and Ellerstrom, M. (2006). Phospholipase-dependent signalling during the AvrRpm1- and AvrRpt2-induced disease resistance responses in Arabidopsis thaliana. The Plant journal : for cell and molecular biology *47*, 947-959. doi: 10.1111/j.1365-313X.2006.02844.x.

Arbona, V., and Gomez-Cadenas, A. (2015). Metabolomics of Disease Resistance in Crops. Current issues in molecular biology *19*, 13-30.

Asai, T., Tena, G., Plotnikova, J., Willmann, M.R., Chiu, W.L., Gomez-Gomez, L., Boller, T., Ausubel, F.M., and Sheen, J. (2002). MAP kinase signalling cascade in Arabidopsis innate immunity. Nature *415*, 977-983. doi: 10.1038/415977a.

Bailey, B. (1995). Purification of a protein from culture filtrates of Fusarium oxysporum that induces ethylene and necrosis in leaves of Erythroxylum coca. Phytopathology *1250-55*.

Bauer, Z., Gomez-Gomez, L., Boller, T., and Felix, G. (2001). Sensitivity of different ecotypes and mutants of Arabidopsis thaliana toward the bacterial elicitor flagellin correlates with the presence of receptor-binding sites. The Journal of biological chemistry *276*, 45669-45676. doi: 10.1074/jbc.M102390200.

Baxter, A., Mittler, R., and Suzuki, N. (2014). ROS as key players in plant stress signalling. Journal of experimental botany *65*, 1229-1240. doi: 10.1093/jxb/ert375.

Beckers, G.J., Jaskiewicz, M., Liu, Y., Underwood, W.R., He, S.Y., Zhang, S., and Conrath, U. (2009). Mitogen-activated protein kinases 3 and 6 are required for full priming of stress responses in Arabidopsis thaliana. The Plant cell *21*, 944-953. doi: 10.1105/tpc.108.062158.

Berr, A., Menard, R., Heitz, T., and Shen, W.H. (2012). Chromatin modification and remodelling: a regulatory landscape for the control of Arabidopsis defence responses upon pathogen attack. Cellular microbiology *14*, 829-839. doi: 10.1111/j.1462-5822.2012.01785.x.

Berriri, S., Garcia, A.V., Frei dit Frey, N., Rozhon, W., Pateyron, S., Leonhardt, N., Montillet, J.L., Leung, J., Hirt, H., and Colcombet, J. (2012). Constitutively active mitogen-activated protein kinase versions reveal functions of Arabidopsis MPK4 in pathogen defense signaling. The Plant cell *24*, 4281-4293. doi: 10.1105/tpc.112.101253.

Bethke, G., Pecher, P., Eschen-Lippold, L., Tsuda, K., Katagiri, F., Glazebrook, J., Scheel, D., and Lee, J. (2012). Activation of the Arabidopsis thaliana mitogen-activated protein kinase MPK11 by the flagellin-derived elicitor peptide, flg22. Molecular plant-microbe interactions : MPMI *25*, 471-480. doi: 10.1094/mpmi-11-11-0281.

Bethke, G., Unthan, T., Uhrig, J.F., Poschl, Y., Gust, A.A., Scheel, D., and Lee, J. (2009). Flg22 regulates the release of an ethylene response factor substrate from MAP kinase 6 in Arabidopsis thaliana via ethylene signaling. Proceedings of the National Academy of Sciences of the United States of America *106*, 8067-8072. doi: 10.1073/pnas.0810206106.

Bhattacharjee, S., Garner, C.M., and Gassmann, W. (2013). New clues in the nucleus: transcriptional reprogramming in effector-triggered immunity. Frontiers in plant science *4*, 364. doi: 10.3389/fpls.2013.00364.

Bhattacharjee, S., Halane, M.K., Kim, S.H., and Gassmann, W. (2011). Pathogen effectors target Arabidopsis EDS1 and alter its interactions with immune regulators. Science (New York, NY) *334*, 1405-1408. doi: 10.1126/science.1211592.

Bigeard, J., Colcombet, J., and Hirt, H. (2015). Signaling mechanisms in pattern-triggered immunity (PTI). Molecular plant *8*, 521-539. doi: 10.1016/j.molp.2014.12.022.

Bohm, H., Albert, I., Fan, L., Reinhard, A., and Nurnberger, T. (2014a). Immune receptor complexes at the plant cell surface. Current opinion in plant biology *20*, 47-54. doi: 10.1016/j.pbi.2014.04.007.

Bohm, H., Albert, I., Oome, S., Raaymakers, T.M., Van den Ackerveken, G., and Nurnberger, T.

(2014b). A conserved peptide pattern from a widespread microbial virulence factor triggers pattern-induced immunity in Arabidopsis. PLoS pathogens *10*, e1004491. doi: 10.1371/journal.ppat.1004491.

Boller, T., and Felix, G. (2009). A renaissance of elicitors: perception of microbe-associated molecular patterns and danger signals by pattern-recognition receptors. Annual review of plant biology *60*, 379-406. doi: 10.1146/annurev.arplant.57.032905.105346.

Boudsocq, M., and Sheen, J. (2013). CDPKs in immune and stress signaling. Trends in plant science *18*, 30-40.

Boudsocq, M., Willmann, M.R., McCormack, M., Lee, H., Shan, L., He, P., Bush, J., Cheng, S.H., and Sheen, J. (2010). Differential innate immune signalling via Ca(2+) sensor protein kinases. Nature *464*, 418-422. doi: 10.1016/j.tplants.2012.08.008.

Boyes, D.C., Nam, J., and Dangl, J.L. (1998). The Arabidopsis thaliana RPM1 disease resistance gene product is a peripheral plasma membrane protein that is degraded coincident with the hypersensitive response. Proceedings of the National Academy of Sciences of the United States of America *95*, 15849-15854.

Brodersen, P., Petersen, M., Bjorn Nielsen, H., Zhu, S., Newman, M.A., Shokat, K.M., Rietz, S., Parker, J., and Mundy, J. (2006). Arabidopsis MAP kinase 4 regulates salicylic acid- and jasmonic acid/ethylene-dependent responses via EDS1 and PAD4. The Plant journal : for cell and molecular biology *47*, 532-546. doi: 10.1111/j.1365-313X.2006.02806.x.

Buell, C.R., Joardar, V., Lindeberg, M., Selengut, J., Paulsen, I.T., Gwinn, M.L., Dodson, R.J., Deboy, R.T., Durkin, A.S., Kolonay, J.F., *et al.* (2003). The complete genome sequence of the Arabidopsis and tomato pathogen Pseudomonas syringae pv. tomato DC3000. Proceedings of the National Academy of Sciences of the United States of America *100*, 10181-10186. doi: 10.1073/pnas.1731982100.

Cai, R., Lewis, J., Yan, S., Liu, H., Clarke, C.R., Campanile, F., Almeida, N.F., Studholme, D.J., Lindeberg, M., Schneider, D., *et al.* (2011). The plant pathogen Pseudomonas syringae pv. tomato is genetically monomorphic and under strong selection to evade tomato immunity. PLoS pathogens *7*, e1002130. doi: 10.1371/journal.ppat.1002130.

Casper-Lindley, C., Dahlbeck, D., Clark, E.T., and Staskawicz, B.J. (2002). Direct biochemical evidence for type III secretion-dependent translocation of the AvrBs2 effector protein into plant cells. Proceedings of the National Academy of Sciences of the United States of America *99*, 8336-8341. doi: 10.1073/pnas.122220299.

Cesari, S., Bernoux, M., Moncuquet, P., Kroj, T., and Dodds, P.N. (2014). A novel conserved mechanism for plant NLR protein pairs: the "integrated decoy" hypothesis. Frontiers in plant science *5*, 606. doi: 10.3389/fpls.2014.00606.

Chang, C., Yu, D., Jiao, J., Jing, S., Schulze-Lefert, P., and Shen, Q.H. (2013). Barley MLA immune receptors directly interfere with antagonistically acting transcription factors to initiate disease resistance signaling. The Plant cell *25*, 1158-1173. doi: 10.1105/tpc.113.109942.

Cheng, Z., Li, J.F., Niu, Y., Zhang, X.C., Woody, O.Z., Xiong, Y., Djonovic, S., Millet, Y., Bush, J., McConkey, B.J., *et al.* (2015). Pathogen-secreted proteases activate a novel plant immune pathway. Nature *521*, 213-216. doi: 10.1038/nature14243.

Chinchilla, D., Zipfel, C., Robatzek, S., Kemmerling, B., Nurnberger, T., Jones, J.D., Felix, G., and Boller, T. (2007). A flagellin-induced complex of the receptor FLS2 and BAK1 initiates plant defence. Nature *448*, 497-500. doi: 10.1038/nature05999.

Chisholm, S.T., Coaker, G., Day, B., and Staskawicz, B.J. (2006). Host-microbe interactions: shaping the evolution of the plant immune response. Cell *124*, 803-814. doi: 10.1016/j.cell.2006.02.008.

Clarke, C.R., Chinchilla, D., Hind, S.R., Taguchi, F., Miki, R., Ichinose, Y., Martin, G.B., Leman, S., Felix, G., and Vinatzer, B.A. (2013). Allelic variation in two distinct Pseudomonas syringae flagellin epitopes modulates the strength of plant immune responses but not bacterial motility. The New phytologist *200*, 847-860. doi: 10.1111/nph.12408.

Colcombet, J., and Hirt, H. (2008). Arabidopsis MAPKs: a complex signalling network involved in multiple biological processes. The Biochemical journal *413*, 217-226. doi: 10.1042/bj20080625.

Cook, D.E., Mesarich, C.H., and Thomma, B.P. (2015). Understanding plant immunity as a surveillance system to detect invasion. Annual review of phytopathology *53*, 541-563. doi: 10.1146/annurev-phyto-080614-120114.

Cui, H., Wang, Y., Xue, L., Chu, J., Yan, C., Fu, J., Chen, M., Innes, R.W., and Zhou, J.M. (2010). Pseudomonas syringae effector protein AvrB perturbs Arabidopsis hormone signaling by activating MAP kinase 4. Cell host and microbe *7*, 164-175. doi: 10.1016/j.chom.2010.01.009.

Dangl, J.L., and Jones, J.D. (2001). Plant pathogens and integrated defence responses to infection. Nature *411*, 826-833. doi: 10.1038/35081161.

de Jonge, R., and Thomma, B.P. (2009). Fungal LysM effectors: extinguishers of host immunity? Trends in microbiology *17*, 151-157. doi: 10.1016/j.tim.2009.01.002.

de Jonge, R., van Esse, H.P., Kombrink, A., Shinya, T., Desaki, Y., Bours, R., van der Krol, S., Shibuya, N., Joosten, M.H., and Thomma, B.P. (2010). Conserved fungal LysM effector Ecp6 prevents chitin-triggered immunity in plants. Science (New York, NY) *329*, 953-955. doi: 10.1126/science.1190859.

Denoux, C., Galletti, R., Mammarella, N., Gopalan, S., Werck, D., De Lorenzo, G., Ferrari, S., Ausubel, F.M., and Dewdney, J. (2008). Activation of defense response pathways by OGs and Flg22 elicitors in Arabidopsis seedlings. Molecular plant *1*, 423-445. doi: 10.1093/mp/ssn019.

Djamei, A., Pitzschke, A., Nakagami, H., Rajh, I., and Hirt, H. (2007). Trojan horse strategy in Agrobacterium transformation: abusing MAPK defense signaling. Science (New York, NY) *318*, 453-456. doi: 10.1126/science.1148110.

Dodds, P.N., and Rathjen, J.P. (2010). Plant immunity: towards an integrated view of plant-pathogen interactions. Nature reviews Genetics *11*, 539-548. doi: 10.1038/nrg2812.

Dowen, R.H., Engel, J.L., Shao, F., Ecker, J.R., and Dixon, J.E. (2009). A family of bacterial cysteine protease type III effectors utilizes acylation-dependent and -independent strategies to localize to plasma membranes. The Journal of biological chemistry *284*, 15867-15879. doi: 10.1074/jbc.M900519200.

Dowen, R.H., Pelizzola, M., Schmitz, R.J., Lister, R., Dowen, J.M., Nery, J.R., Dixon, J.E., and Ecker, J.R. (2012). Widespread dynamic DNA methylation in response to biotic stress. Proceedings of the National Academy of Sciences of the United States of America *109*, E2183-2191. doi: 10.1073/pnas.1209329109.

Dubiella, U., Seybold, H., Durian, G., Komander, E., Lassig, R., Witte, C.P., Schulze, W.X., and Romeis, T. (2013). Calcium-dependent protein kinase/NADPH oxidase activation circuit is required for rapid defense signal propagation. Proceedings of the National Academy of Sciences of the United States of America *110*, 8744-8749. doi: 10.1073/pnas.1221294110.

Engelhardt, S., Lee, J., Gabler, Y., Kemmerling, B., Haapalainen, M.L., Li, C.M., Wei, Z., Keller, H., Joosten, M., Taira, S., *et al.* (2009). Separable roles of the Pseudomonas syringae pv. phaseolicola accessory protein HrpZ1 in ion-conducting pore formation and activation of plant immunity. The Plant journal : for cell and molecular biology *57*, 706-717. doi: 10.1111/j.1365-313X.2008.03723.x.

Espinosa, A., Guo, M., Tam, V.C., Fu, Z.Q., and Alfano, J.R. (2003). The Pseudomonas syringae type III-secreted protein HopPtoD2 possesses protein tyrosine phosphatase activity and suppresses programmed cell death in plants. Molecular microbiology *49*, 377-387.

Eulgem, T., and Somssich, I.E. (2007). Networks of WRKY transcription factors in defense signaling. Current opinion in plant biology *10*, 366-371. doi: 10.1016/j.pbi.2007.04.020.

Farmer, P.K., and Choi, J.H. (1999). Calcium and phospholipid activation of a recombinant calcium-dependent protein kinase (DcCPK1) from carrot (Daucus carota L.). Biochimica et biophysica acta *1434*, 6-17.

Felix, G., Duran, J.D., Volko, S., and Boller, T. (1999). Plants have a sensitive perception system for the most conserved domain of bacterial flagellin. The Plant journal : for cell and molecular biology *18*, 265-276.

Feng, F., and Zhou, J.M. (2012). Plant-bacterial pathogen interactions mediated by type III effectors. Current opinion in plant biology *15*, 469-476. doi: 10.1016/j.pbi.2012.03.004.

Frei dit Frey, N., Garcia, A.V., Bigeard, J., Zaag, R., Bueso, E., Garmier, M., Pateyron, S., de Tauzia-Moreau, M.L., Brunaud, V., Balzergue, S., *et al.* (2014). Functional analysis of Arabidopsis immune-related MAPKs uncovers a role for MPK3 as negative regulator of inducible defences. Genome biology *15*, R87. doi: 10.1186/gb-2014-15-6-r87.

Frias, M., Brito, N., Gonzalez, M., and Gonzalez, C. (2014). The phytotoxic activity of the cerato-platanin BcSpl1 resides in a two-peptide motif on the protein surface. Molecular plant pathology *15*, 342-351. doi: 10.1111/mpp.12097.

Frias, M., Gonzalez, C., and Brito, N. (2011). BcSpl1, a cerato-platanin family protein, contributes to Botrytis cinerea virulence and elicits the hypersensitive response in the host. The New phytologist *192*, 483-495. doi: 10.1111/j.1469-8137.2011.03802.x.

Gao, M., Liu, J., Bi, D., Zhang, Z., Cheng, F., Chen, S., and Zhang, Y. (2008). MEKK1, MKK1/MKK2 and MPK4 function together in a mitogen-activated protein kinase cascade to regulate innate immunity in plants. Cell research *18*, 1190-1198. doi: 10.1038/cr.2008.300.

Gao, X., Chen, X., Lin, W., Chen, S., Lu, D., Niu, Y., Li, L., Cheng, C., McCormack, M., Sheen, J., *et al.* (2013). Bifurcation of Arabidopsis NLR immune signaling via Ca(2)(+)-dependent protein kinases. PLoS pathogens *9*, e1003127. doi: 10.1371/journal.ppat.1003127.

Gao, Z., Chung, E.H., Eitas, T.K., and Dangl, J.L. (2011). Plant intracellular innate immune receptor Resistance to Pseudomonas syringae pv. maculicola 1 (RPM1) is activated at, and functions on, the plasma membrane. Proceedings of the National Academy of Sciences of the United States of America *108*, 7619-7624. doi: 10.1073/pnas.1104410108.

Garcia, A.V., Blanvillain-Baufume, S., Huibers, R.P., Wiermer, M., Li, G., Gobbato, E., Rietz, S., and Parker, J.E. (2010). Balanced nuclear and cytoplasmic activities of EDS1 are required for a complete plant innate immune response. PLoS pathogens *6*, e1000970. doi: 10.1371/journal.ppat.1000970.

Gassmann, W., Hinsch, M.E., and Staskawicz, B.J. (1999). The Arabidopsis RPS4 bacterial-resistance gene is a member of the TIR-NBS-LRR family of disease-resistance genes. The Plant journal : for cell and molecular biology *20*, 265-277.

Gatz, C. (2013). From pioneers to team players: TGA transcription factors provide a molecular link between different stress pathways. Molecular plant-microbe interactions : MPMI *26*, 151-159. doi: 10.1094/mpmi-04-12-0078-ia.

Gijzen, M., and Nurnberger, T. (2006). Nep1-like proteins from plant pathogens: recruitment and diversification of the NPP1 domain across taxa. Phytochemistry *67*, 1800-1807. doi: 10.1016/j.phytochem.2005.12.008.

Gomez-Gomez, L., and Boller, T. (2000). FLS2: an LRR receptor-like kinase involved in the perception of the bacterial elicitor flagellin in Arabidopsis. Molecular cell *5*, 1003-1011.

Gomez-Gomez, L., and Boller, T. (2002). Flagellin perception: a paradigm for innate immunity. Trends in plant science *7*, 251-256.

Grant, M.R., Godiard, L., Straube, E., Ashfield, T., Lewald, J., Sattler, A., Innes, R.W., and Dangl, J.L. (1995). Structure of the Arabidopsis RPM1 gene enabling dual specificity disease resistance. Science (New York, NY) *269*, 843-846.

Gross, F., Durner, J., and Gaupels, F. (2013). Nitric oxide, antioxidants and prooxidants in plant defence responses. Frontiers in plant science *4*, 419. doi: 10.3389/fpls.2013.00419.

Gutterson, N., and Reuber, T.L. (2004). Regulation of disease resistance pathways by AP2/ERF transcription factors. Current opinion in plant biology *7*, 465-471. doi: 10.1016/j.pbi.2004.04.007.

Hadiarto, T., Nanmori, T., Matsuoka, D., Iwasaki, T., Sato, K., Fukami, Y., Azuma, T., and Yasuda, T. (2006). Activation of Arabidopsis MAPK kinase kinase (AtMEKK1) and induction of AtMEKK1-AtMEK1 pathway by wounding. Planta *223*, 708-713. doi: 10.1007/s00425-005-0126-7.

Han, L., Li, G.J., Yang, K.Y., Mao, G., Wang, R., Liu, Y., and Zhang, S. (2010). Mitogen-activated protein kinase 3 and 6 regulate Botrytis cinerea-induced ethylene production in Arabidopsis. The Plant journal : for cell and molecular biology *64*, 114-127. doi: 10.1111/j.1365-313X.2010.04318.x.

Hann, D.R., Gimenez-Ibanez, S., and Rathjen, J.P. (2010). Bacterial virulence effectors and their activities. Current opinion in plant biology *13*, 388-393. doi: 10.1016/j.pbi.2010.04.003.

Heidrich, K., Wirthmueller, L., Tasset, C., Pouzet, C., Deslandes, L., and Parker, J.E. (2011). Arabidopsis EDS1 connects pathogen effector recognition to cell compartment-specific immune responses. Science (New York, NY) *334*, 1401-1404. doi: 10.1126/science.1211641.

Hogenhout, S.A., Van der Hoorn, R.A., Terauchi, R., and Kamoun, S. (2009). Emerging concepts in effector biology of plant-associated organisms. Molecular plant-microbe interactions : MPMI *22*, 115-122. doi: 10.1094/mpmi-22-2-0115.

Huang, S., Chen, X., Zhong, X., Li, M., Ao, K., Huang, J., and Li, X. (2016). Plant TRAF Proteins Regulate NLR Immune Receptor Turnover. Cell host and microbe *19*, 204-215. doi: 10.1016/j.chom.2016.01.005.

Huang, Y., Li, H., Gupta, R., Morris, P.C., Luan, S., and Kieber, J.J. (2000). ATMPK4, an Arabidopsis homolog of mitogen-activated protein kinase, is activated in vitro by AtMEK1 through threonine phosphorylation. Plant physiology *122*, 1301-1310.

Ichimura, K., Casais, C., Peck, S.C., Shinozaki, K., and Shirasu, K. (2006). MEKK1 is required for MPK4 activation and regulates tissue-specific and temperature-dependent cell death in Arabidopsis. The Journal of biological chemistry *281*, 36969-36976. doi: 10.1074/jbc.M605319200.

Ichimura, K., Mizoguchi, T., Irie, K., Morris, P., Giraudat, J., Matsumoto, K., and Shinozaki, K. (1998). Isolation of ATMEKK1 (a MAP kinase kinase kinase)-interacting proteins and analysis of a MAP kinase cascade in Arabidopsis. Biochemical and biophysical research communications *253*, 532-543. doi: 10.1006/bbrc.1998.9796.

Inoue, H., Hayashi, N., Matsushita, A., Xinqiong, L., Nakayama, A., Sugano, S., Jiang, C.J., and Takatsuji, H. (2013). Blast resistance of CC-NB-LRR protein Pb1 is mediated by WRKY45 through protein-protein interaction. Proceedings of the National Academy of Sciences of the United States of America *110*, 9577-9582. doi: 10.1073/pnas.1222155110.

Jacob, F., Vernaldi, S., and Maekawa, T. (2013). Evolution and Conservation of Plant NLR Functions. Frontiers in immunology *4*, 297. doi: 10.3389/fimmu.2013.00297.

Jeworutzki, E., Roelfsema, M.R., Anschutz, U., Krol, E., Elzenga, J.T., Felix, G., Boller, T., Hedrich, R., and Becker, D. (2010). Early signaling through the Arabidopsis pattern recognition receptors FLS2 and EFR involves Ca-associated opening of plasma membrane anion channels. The Plant journal : for cell and molecular biology *62*, 367-378. doi: 10.1111/j.1365-313X.2010.04155.x.

Kadota, Y., Shirasu, K., and Guerois, R. (2010). NLR sensors meet at the SGT1-HSP90 crossroad. Trends in biochemical sciences *35*, 199-207. doi: 10.1016/j.tibs.2009.12.005.

Kadota, Y., Sklenar, J., Derbyshire, P., Stransfeld, L., Asai, S., Ntoukakis, V., Jones, J.D., Shirasu, K., Menke, F., Jones, A., *et al.* (2014). Direct regulation of the NADPH oxidase RBOHD by the PRR-associated kinase BIK1 during plant immunity. Molecular cell *54*, 43-55. doi: 10.1016/j.molcel.2014.02.021.

Kamiyoshihara, Y., Iwata, M., Fukaya, T., Tatsuki, M., and Mori, H. (2010). Turnover of LeACS2, a wound-inducible 1-aminocyclopropane-1-carboxylic acid synthase in tomato, is regulated by phosphorylation/dephosphorylation. The Plant journal : for cell and molecular biology *64*, 140-150. doi: 10.1111/j.1365-313X.2010.04316.x.

Kim, C.Y., Liu, Y., Thorne, E.T., Yang, H., Fukushige, H., Gassmann, W., Hildebrand, D., Sharp, R.E., and Zhang, S. (2003). Activation of a stress-responsive mitogen-activated protein kinase cascade induces the biosynthesis of ethylene in plants. The Plant cell *15*, 2707-2718. doi: 10.1105/tpc.011411.

Kim, H., O'Connell, R., Maekawa-Yoshikawa, M., Uemura, T., Neumann, U., and Schulze-Lefert, P. (2014a). The powdery mildew resistance protein RPW8.2 is carried on VAMP721/722 vesicles to the extrahaustorial membrane of haustorial complexes. The Plant journal : for cell and molecular biology *79*, 835-847. doi: 10.1111/tpj.12591.

Kim, S.H., Son, G.H., Bhattacharjee, S., Kim, H.J., Nam, J.C., Nguyen, P.D., Hong, J.C., and Gassmann, W. (2014b). The Arabidopsis immune adaptor SRFR1 interacts with TCP transcription factors that redundantly contribute to effector-triggered immunity. The Plant journal : for cell and molecular biology *78*, 978-989. doi: 10.1111/tpj.12527.

Kong, Q., Qu, N., Gao, M., Zhang, Z., Ding, X., Yang, F., Li, Y., Dong, O.X., Chen, S., Li, X., *et al.* (2012). The MEKK1-MKK1/MKK2-MPK4 kinase cascade negatively regulates immunity mediated by a mitogen-activated protein kinase kinase kinase in Arabidopsis. The Plant cell *24*, 2225-2236. doi: 10.1105/tpc.112.097253.

Kroj, T., Chanclud, E., Michel-Romiti, C., Grand, X., and Morel, J.B. (2016). Integration of decoy domains derived from protein targets of pathogen effectors into plant immune receptors is widespread. The New phytologist. doi: 10.1111/nph.13869.

Lassowskat, I., Bottcher, C., Eschen-Lippold, L., Scheel, D., and Lee, J. (2014). Sustained mitogen-activated protein kinase activation reprograms defense metabolism and phosphoprotein profile in Arabidopsis thaliana. Frontiers in plant science *5*, 554. doi: 10.3389/fpls.2014.00554.

Liu, Y., and Zhang, S. (2004). Phosphorylation of 1-aminocyclopropane-1-carboxylic acid synthase by MPK6, a stress-responsive mitogen-activated protein kinase, induces ethylene biosynthesis in Arabidopsis. The Plant cell *16*, 3386-3399. doi: 10.1105/tpc.104.026609.

Liu, Z., Wu, Y., Yang, F., Zhang, Y., Chen, S., Xie, Q., Tian, X., and Zhou, J.M. (2013). BIK1 interacts with PEPRs to mediate ethylene-induced immunity. Proceedings of the National Academy of Sciences of the United States of America *110*, 6205-6210. doi: 10.1073/pnas.1215543110.

Lohou, D., Lonjon, F., Genin, S., and Vailleau, F. (2013). Type III chaperones and Co in bacterial plant pathogens: a set of specialized bodyguards mediating effector delivery. Frontiers in plant science *4*, 435. doi: 10.3389/fpls.2013.00435.

Longhi, S., and Cambillau, C. (1999). Structure-activity of cutinase, a small lipolytic enzyme. Biochimica et biophysica acta *1441*, 185-196.

Lu, D., Wu, S., Gao, X., Zhang, Y., Shan, L., and He, P. (2010). A receptor-like cytoplasmic kinase, BIK1, associates with a flagellin receptor complex to initiate plant innate immunity. Proceedings of the National Academy of Sciences of the United States of America *107*, 496-501. doi: 10.1073/pnas.0909705107.

Ma, K.W., Flores, C., and Ma, W. (2011). Chromatin configuration as a battlefield in plant-bacteria interactions. Plant physiology *157*, 535-543. doi: 10.1104/pp.111.182295.

Macho, A.P., Schwessinger, B., Ntoukakis, V., Brutus, A., Segonzac, C., Roy, S., Kadota, Y., Oh, M.H., Sklenar, J., Derbyshire, P., *et al.* (2014). A bacterial tyrosine phosphatase inhibits plant pattern recognition receptor activation. Science (New York, NY) *343*, 1509-1512. doi: 10.1126/science.1248849.

Macho, A.P., and Zipfel, C. (2014). Plant PRRs and the activation of innate immune signaling. Molecular cell *54*, 263-272. doi: 10.1016/j.molcel.2014.03.028.

Maekawa, T., Kufer, T.A., and Schulze-Lefert, P. (2011). NLR functions in plant and animal immune systems: so far and yet so close. Nature immunology *12*, 817-826. doi: 10.1038/ni.2083.

Matsuoka, D., Nanmori, T., Sato, K., Fukami, Y., Kikkawa, U., and Yasuda, T. (2002). Activation of AtMEK1, an Arabidopsis mitogen-activated protein kinase kinase, in vitro and in vivo: analysis of active mutants expressed in E. coli and generation of the active form in stress response in seedlings. The Plant journal : for cell and molecular biology *29*, 637-647.

Melotto, M., Underwood, W., and He, S.Y. (2008). Role of stomata in plant innate immunity and foliar bacterial diseases. Annual review of phytopathology *46*, 101-122. doi: 10.1146/annurev.phyto.121107.104959.

Mendgen, K., Hahn, M., and Deising, H. (1996). Morphogenesis and mechanisms of penetration by plant pathogenic fungi. Annual review of phytopathology *34*, 367-386. doi: 10.1146/annurev.phyto.34.1.367.

Meng, X., and Zhang, S. (2013). MAPK cascades in plant disease resistance signaling. Annual review of phytopathology *51*, 245-266. doi: 10.1146/annurev-phyto-082712-102314.

Mizoguchi, T., Ichimura, K., Irie, K., Morris, P., Giraudat, J., Matsumoto, K., and Shinozaki, K. (1998). Identification of a possible MAP kinase cascade in Arabidopsis thaliana based on pairwise yeast two-hybrid analysis and functional complementation tests of yeast mutants. FEBS letters *437*, 56-60.

Monaghan, J., Matschi, S., Romeis, T., and Zipfel, C. (2015). The calcium-dependent protein kinase CPK28 negatively regulates the BIK1-mediated PAMP-induced calcium burst. Plant signaling and behavior *10*, e1018497. doi: 10.1080/15592324.2015.1018497.

Monaghan, J., Matschi, S., Shorinola, O., Rovenich, H., Matei, A., Segonzac, C., Malinovsky, F.G., Rathjen, J.P., MacLean, D., Romeis, T., *et al.* (2014). The calcium-dependent protein kinase CPK28 buffers plant immunity and regulates BIK1 turnover. Cell host and microbe *16*, 605-615. doi: 10.1016/j.chom.2014.10.007.

Monaghan, J., and Zipfel, C. (2012). Plant pattern recognition receptor complexes at the plasma membrane. Current opinion in plant biology *15*, 349-357. doi: 10.1016/j.pbi.2012.05.006.

Moore, J.W., Loake, G.J., and Spoel, S.H. (2011). Transcription dynamics in plant immunity. The Plant cell *23*, 2809-2820. doi: 10.1105/tpc.111.087346.

Mudgett, M.B., and Staskawicz, B.J. (1999). Characterization of the Pseudomonas syringae pv. tomato AvrRpt2 protein: demonstration of secretion and processing during bacterial pathogenesis. Molecular microbiology *32*, 927-941.

Mukhtar, M.S., Carvunis, A.R., Dreze, M., Epple, P., Steinbrenner, J., Moore, J., Tasan, M., Galli, M., Hao, T., Nishimura, M.T., *et al.* (2011). Independently evolved virulence effectors converge onto hubs in a plant immune system network. Science (New York, NY) *333*, 596-601.

Nakagami, H., Soukupova, H., Schikora, A., Zarsky, V., and Hirt, H. (2006). A Mitogen-activated protein kinase kinase kinase mediates reactive oxygen species homeostasis in Arabidopsis. The Journal of biological chemistry *281*, 38697-38704. doi: 10.1126/science.1203659. oi: 10.1074/jbc.M605293200

Nakano, M., Nishihara, M., Yoshioka, H., Takahashi, H., Sawasaki, T., Ohnishi, K., Hikichi, Y., and Kiba, A. (2013). Suppression of DS1 phosphatidic acid phosphatase confirms resistance to Ralstonia solanacearum in Nicotiana benthamiana. PloS one *8*, e75124. doi: 10.1371/journal.pone.0075124.

Narusaka, M., Shirasu, K., Noutoshi, Y., Kubo, Y., Shiraishi, T., Iwabuchi, M., and Narusaka, Y. (2009). RRS1 and RPS4 provide a dual Resistance-gene system against fungal and bacterial pathogens. The Plant journal : for cell and molecular biology *60*, 218-226. doi: 10.1111/j.1365-313X.2009.03949.x.

Nitta, Y., Ding, P., and Zhang, Y. (2014). Identification of additional MAP kinases activated upon PAMP treatment. Plant signaling and behavior *9*, e976155. doi: 10.4161/15592324.2014.976155.

Nomura, H., Komori, T., Uemura, S., Kanda, Y., Shimotani, K., Nakai, K., Furuichi, T., Takebayashi, K., Sugimoto, T., Sano, S., *et al.* (2012). Chloroplast-mediated activation of plant immune signalling in Arabidopsis. Nature communications *3*, 926. doi: 10.1038/ncomms1926.

Nuhse, T.S., Bottrill, A.R., Jones, A.M., and Peck, S.C. (2007). Quantitative phosphoproteomic analysis of plasma membrane proteins reveals regulatory mechanisms of plant innate immune responses. The Plant journal : for cell and molecular biology *51*, 931-940. doi: 10.1111/j.1365-313X.2007.03192.x.

Nuhse, T.S., Peck, S.C., Hirt, H., and Boller, T. (2000). Microbial elicitors induce activation and dual phosphorylation of the Arabidopsis thaliana MAPK 6. The Journal of biological chemistry *275*, 7521-7526.

Nuruzzaman, M., Sharoni, A.M., and Kikuchi, S. (2013). Roles of NAC transcription factors in the regulation of biotic and abiotic stress responses in plants. Frontiers in microbiology *4*, 248. doi: 10.3389/fmicb.2013.00248.

Ogasawara, Y., Kaya, H., Hiraoka, G., Yumoto, F., Kimura, S., Kadota, Y., Hishinuma, H., Senzaki, E., Yamagoe, S., Nagata, K., *et al.* (2008). Synergistic activation of the Arabidopsis NADPH oxidase AtrbohD by Ca2+ and phosphorylation. The Journal of biological chemistry *283*, 8885-8892. doi: 10.1074/jbc.M708106200.

Oh, C.S., and Martin, G.B. (2011). Tomato 14-3-3 protein TFT7 interacts with a MAP kinase kinase to regulate immunity-associated programmed cell death mediated by diverse disease resistance proteins. The Journal of biological chemistry *286*, 14129-14136. doi: 10.1074/jbc.M111.225086.

Oh, C.S., Pedley, K.F., and Martin, G.B. (2010). Tomato 14-3-3 protein 7 positively regulates immunity-associated programmed cell death by enhancing protein abundance and signaling ability of MAPKKK {alpha}. The Plant cell *22*, 260-272. doi: 10.1105/tpc.109.070664.

Okazaki, Y., and Saito, K. (2014). Roles of lipids as signaling molecules and mitigators during stress response in plants. The Plant journal : for cell and molecular biology *79*, 584-596. doi: 10.1111/tpj.12556.

Oome, S., Raaymakers, T.M., Cabral, A., Samwel, S., Bohm, H., Albert, I., Nurnberger, T., and Van den Ackerveken, G. (2014). Nep1-like proteins from three kingdoms of life act as a microbe-associated molecular pattern in Arabidopsis. Proceedings of the National Academy of Sciences of the United States of America *111*, 16955-16960. doi: 10.1073/pnas.1410031111.

Osbourn, A.E. (1996). Preformed Antimicrobial Compounds and Plant Defense against Fungal Attack. The Plant cell *8*, 1821-1831. doi: 10.1105/tpc.8.10.1821.

Ottmann, C., Luberacki, B., Kufner, I., Koch, W., Brunner, F., Weyand, M., Mattinen, L., Pirhonen, M., Anderluh, G., Seitz, H.U., *et al.* (2009). A common toxin fold mediates microbial attack and plant defense. Proceedings of the National Academy of Sciences of the United States of America *106*, 10359-10364. doi: 10.1073/pnas.0902362106.

Pandey, S.P., and Somssich, I.E. (2009). The role of WRKY transcription factors in plant immunity. Plant physiology *150*, 1648-1655. doi: 10.1104/pp.109.138990.

Petersen, M., Brodersen, P., Naested, H., Andreasson, E., Lindhart, U., Johansen, B., Nielsen, H.B., Lacy, M., Austin, M.J., Parker, J.E., *et al.* (2000). Arabidopsis map kinase 4 negatively regulates systemic acquired resistance. Cell *103*, 1111-1120.

Pfund, C., Tans-Kersten, J., Dunning, F.M., Alonso, J.M., Ecker, J.R., Allen, C., and Bent, A.F. (2004). Flagellin is not a major defense elicitor in Ralstonia solanacearum cells or extracts applied to Arabidopsis thaliana. Molecular plant-microbe interactions : MPMI *17*, 696-706. doi: 10.1094/mpmi.2004.17.6.696.

Pitzschke, A., Schikora, A., and Hirt, H. (2009). MAPK cascade signalling networks in plant defence. Current opinion in plant biology *12*, 421-426. doi: 10.1016/j.pbi.2009.06.008.

Puranik, S., Sahu, P.P., Srivastava, P.S., and Prasad, M. (2012). NAC proteins: regulation and role in stress tolerance. Trends in plant science *17*, 369-381. doi: 10.1016/j.tplants.2012.02.004.

Qi, D., and Innes, R.W. (2013). Recent Advances in Plant NLR Structure, Function, Localization, and Signaling. Frontiers in immunology *4*, 348. doi: 10.3389/fimmu.2013.00348.

Qi, Y., Tsuda, K., Glazebrook, J., and Katagiri, F. (2011). Physical association of pattern-triggered immunity (PTI) and effector-triggered immunity (ETI) immune receptors in Arabidopsis. Molecular plant pathology *12*, 702-708. doi: 10.1111/j.1364-3703.2010.00704.x.

Qiu, J.L., Zhou, L., Yun, B.W., Nielsen, H.B., Fiil, B.K., Petersen, K., Mackinlay, J., Loake, G.J., Mundy, J., and Morris, P.C. (2008). Arabidopsis mitogen-activated protein kinase kinases MKK1 and MKK2 have overlapping functions in defense signaling mediated by MEKK1, MPK4, and MKS1. Plant physiology *148*, 212-222. doi: 10.1104/pp.108.120006.

Raho, N., Ramirez, L., Lanteri, M.L., Gonorazky, G., Lamattina, L., ten Have, A., and Laxalt, A.M. (2011). Phosphatidic acid production in chitosan-elicited tomato cells, via both phospholipase D and phospholipase C/diacylglycerol kinase, requires nitric oxide. Journal of plant physiology *168*, 534-539. doi: 10.1016/j.jplph.2010.09.004.

Ranf, S., Eschen-Lippold, L., Pecher, P., Lee, J., and Scheel, D. (2011). Interplay between calcium signalling and early signalling elements during defence responses to microbe- or damage-associated molecular patterns. The Plant journal : for cell and molecular biology *68*, 100-113. doi: 10.1111/j.1365-313X.2011.04671.x.

Rasmussen, M.W., Roux, M., Petersen, M., and Mundy, J. (2012). MAP Kinase Cascades in Arabidopsis Innate Immunity. Frontiers in plant science *3*, 169. doi: 10.3389/fpls.2012.00169.

Ren, D., Yang, H., and Zhang, S. (2002). Cell death mediated by MAPK is associated with hydrogen peroxide production in Arabidopsis. The Journal of biological chemistry *277*, 559-565. doi: 10.1074/jbc.M109495200.

Robatzek, S., Bittel, P., Chinchilla, D., Kochner, P., Felix, G., Shiu, S.H., and Boller, T. (2007). Molecular identification and characterization of the tomato flagellin receptor LeFLS2, an orthologue of Arabidopsis FLS2 exhibiting characteristically different perception specificities. Plant molecular biology *64*, 539-547. doi: 10.1007/s11103-007-9173-8.

Rodriguez, M.C., Petersen, M., and Mundy, J. (2010). Mitogen-activated protein kinase signaling in plants. Annual review of plant biology *61*, 621-649. doi: 10.1146/annurev-arplant-042809-112252.

Romeis, T., and Herde, M. (2014). From local to global: CDPKs in systemic defense signaling upon microbial and herbivore attack. Current opinion in plant biology *20*, 1-10. doi: 10.1016/j.pbi.2014.03.002.

Rossi, M., Goggin, F.L., Milligan, S.B., Kaloshian, I., Ullman, D.E., and Williamson, V.M. (1998). The nematode resistance gene Mi of tomato confers resistance against the potato aphid. Proceedings of the National Academy of Sciences of the United States of America *95*, 9750-9754.

Roux, M., Schwessinger, B., Albrecht, C., Chinchilla, D., Jones, A., Holton, N., Malinovsky, F.G., Tor, M., de Vries, S., and Zipfel, C. (2011). The Arabidopsis leucine-rich repeat receptor-like kinases BAK1/SERK3 and BKK1/SERK4 are required for innate immunity to hemibiotrophic and biotrophic pathogens. The Plant cell *23*, 2440-2455. doi: 10.1105/tpc.111.084301.

Sarris, P.F., Cevik, V., Dagdas, G., Jones, J.D., and Krasileva, K.V. (2016). Comparative analysis of plant immune receptor architectures uncovers host proteins likely targeted by pathogens. BMC biology *14*, 8. doi: 10.1186/s12915-016-0228-7.

Schechter, L.M., Roberts, K.A., Jamir, Y., Alfano, J.R., and Collmer, A. (2004). Pseudomonas syringae type III secretion system targeting signals and novel effectors studied with a Cya translocation reporter. Journal of bacteriology *186*, 543-555.

Schulz, P., Herde, M., and Romeis, T. (2013). Calcium-dependent protein kinases: hubs in plant stress signaling and development. Plant physiology *163*, 523-530. doi: 10.1104/pp.113.222539.

Schulze, B., Mentzel, T., Jehle, A.K., Mueller, K., Beeler, S., Boller, T., Felix, G., and Chinchilla, D. (2010). Rapid heteromerization and phosphorylation of ligand-activated plant transmembrane receptors and their associated kinase BAK1. The Journal of biological chemistry *285*, 9444-9451. doi: 10.1074/jbc.M109.096842.

Schwessinger, B., and Ronald, P.C. (2012). Plant innate immunity: perception of conserved microbial signatures. Annual review of plant biology *63*, 451-482. doi: 10.1146/annurev-arplant-042811-105518.

Segonzac, C., and Zipfel, C. (2011). Activation of plant pattern-recognition receptors by bacteria. Current opinion in microbiology *14*, 54-61. doi: 10.1016/j.mib.2010.12.005.

Seo, S., Katou, S., Seto, H., Gomi, K., and Ohashi, Y. (2007). The mitogen-activated protein kinases WIPK and SIPK regulate the levels of jasmonic and salicylic acids in wounded tobacco plants. The Plant journal : for cell and molecular biology *49*, 899-909. doi: 10.1111/j.1365-313X.2006.03003.x.

Shirasu, K. (2009). The HSP90-SGT1 chaperone complex for NLR immune sensors. Annual review of plant biology *60*, 139-164. doi: 10.1146/annurev.arplant.59.032607.092906.

Smith, J.M., and Heese, A. (2014). Rapid bioassay to measure early reactive oxygen species production in Arabidopsis leave tissue in response to living Pseudomonas syringae. Plant methods *10*, 6. doi: 10.1186/1746-4811-10-6.

Somerville, C., Bauer, S., Brininstool, G., Facette, M., Hamann, T., Milne, J., Osborne, E., Paredez, A., Persson, S., Raab, T., *et al.* (2004). Toward a systems approach to understanding plant cell walls. Science (New York, NY) *306*, 2206-2211. doi: 10.1126/science.1102765.

Su, S.H., Bush, S.M., Zaman, N., Stecker, K., Sussman, M.R., and Krysan, P. (2013). Deletion of a tandem gene family in Arabidopsis: increased MEKK2 abundance triggers autoimmunity when the MEKK1-MKK1/2-MPK4 signaling cascade is disrupted. The Plant cell *25*, 1895-1910. doi: 10.1105/tpc.113.112102.

Suarez-Rodriguez, M.C., Adams-Phillips, L., Liu, Y., Wang, H., Su, S.H., Jester, P.J., Zhang, S., Bent, A.F., and Krysan, P.J. (2007). MEKK1 is required for flg22-induced MPK4 activation in Arabidopsis plants. Plant physiology *143*, 661-669. doi: 10.1104/pp.106.091389.

Sun, W., Dunning, F.M., Pfund, C., Weingarten, R., and Bent, A.F. (2006). Within-species flagellin polymorphism in Xanthomonas campestris pv campestris and its impact on elicitation of Arabidopsis FLAGELLIN SENSING2-dependent defenses. The Plant cell *18*, 764-779. doi: 10.1105/tpc.105.037648.

Szczegielniak, J., Klimecka, M., Liwosz, A., Ciesielski, A., Kaczanowski, S., Dobrowolska, G., Harmon, A.C., and Muszynska, G. (2005). A wound-responsive and phospholipid-regulated maize calcium-dependent protein kinase. Plant physiology *139*, 1970-1983. doi: 10.1104/pp. 105.066472.

Tada, Y., Spoel, S.H., Pajerowska-Mukhtar, K., Mou, Z., Song, J., Wang, C., Zuo, J., and Dong, X. (2008). Plant immunity requires conformational changes [corrected] of NPR1 via S-nitrosylation and thioredoxins. Science (New York, NY) *321*, 952-956. doi: 10.1126/science.1156970.

Takahashi, F., Yoshida, R., Ichimura, K., Mizoguchi, T., Seo, S., Yonezawa, M., Maruyama, K., Yamaguchi-Shinozaki, K., and Shinozaki, K. (2007). The mitogen-activated protein kinase cascade MKK3-MPK6 is an important part of the jasmonate signal transduction pathway in Arabidopsis. The Plant cell *19*, 805-818. doi: 10.1105/tpc.106.046581.

Takken, F.L., and Goverse, A. (2012). How to build a pathogen detector: structural basis of NB-LRR function. Current opinion in plant biology *15*, 375-384. doi: 10.1016/j.pbi.2012.05.001.

Tao, Y., Xie, Z., Chen, W., Glazebrook, J., Chang, H.S., Han, B., Zhu, T., Zou, G., and Katagiri, F. (2003). Quantitative nature of Arabidopsis responses during compatible and incompatible interactions with the bacterial pathogen Pseudomonas syringae. The Plant cell *15*, 317-330.

Teige, M., Scheikl, E., Eulgem, T., Doczi, R., Ichimura, K., Shinozaki, K., Dangl, J.L., and Hirt, H. (2004). The MKK2 pathway mediates cold and salt stress signaling in Arabidopsis. Molecular cell *15*, 141-152. doi: 10.1016/j.molcel.2004.06.023.

Teper, D., Salomon, D., Sunitha, S., Kim, J.G., Mudgett, M.B., and Sessa, G. (2014). Xanthomonas euvesicatoria type III effector XopQ interacts with tomato and pepper 14-3-3 isoforms to suppress effector-triggered immunity. The Plant journal : for cell and molecular biology *77*, 297-309. doi: 10.1111/tpj.12391.

Testerink, C., Larsen, P.B., McLoughlin, F., van der Does, D., van Himbergen, J.A., and Munnik, T. (2008). PA, a stress-induced short cut to switch-on ethylene signalling by switching-off CTR1? Plant signaling and behavior *3*, 681-683.

Testerink, C., Larsen, P.B., van der Does, D., van Himbergen, J.A., and Munnik, T. (2007). Phosphatidic acid binds to and inhibits the activity of Arabidopsis CTR1. Journal of experimental botany *58*, 3905-3914. doi: 10.1093/jxb/erm243.

Thomma, B.P., Nurnberger, T., and Joosten, M.H. (2011). Of PAMPs and effectors: the blurred PTI-ETI dichotomy. The Plant cell *23*, 4-15. doi: 10.1105/tpc.110.082602.

Trda, L., Fernandez, O., Boutrot, F., Heloir, M.C., Kelloniemi, J., Daire, X., Adrian, M., Clement, C., Zipfel, C., Dorey, S., *et al.* (2014). The grapevine flagellin receptor VvFLS2 differentially recognizes flagellin-derived epitopes from the endophytic growth-promoting bacterium Burkholderia phytofirmans and plant pathogenic bacteria. The New phytologist *201*, 1371-1384. doi: 10.1111/ nph.12592.

Tsuda, K., and Katagiri, F. (2010). Comparing signaling mechanisms engaged in pattern-triggered and effector-triggered immunity. Current opinion in plant biology *13*, 459-465. doi: 10.1016/ j.pbi.2010.04.006.

Tsuda, K., Mine, A., Bethke, G., Igarashi, D., Botanga, C.J., Tsuda, Y., Glazebrook, J., Sato, M., and Katagiri, F. (2013). Dual regulation of gene expression mediated by extended MAPK activation and salicylic acid contributes to robust innate immunity in Arabidopsis thaliana. PLoS genetics *9*, e1004015. doi: 10.1371/journal.pgen. 1004015.

Tsuda, K., Sato, M., Stoddard, T., Glazebrook, J., and Katagiri, F. (2009). Network properties of robust immunity in plants. PLoS genetics *5*, e1000772. doi: 10.1371/journal.pgen.1000772.

Tzfira, T., Vaidya, M., and Citovsky, V. (2004). Involvement of targeted proteolysis in plant genetic transformation by Agrobacterium. Nature *431*, 87-92. doi: 10.1038/nature02857.

Van der Biezen, E.A., and Jones, J.D. (1998). Plant disease-resistance proteins and the gene-for-gene concept. Trends in biochemical sciences *23*, 454-456.

van der Hoorn, R.A.L., and Kamoun, S. (2008). From Guard to Decoy: A New Model for

Perception of Plant Pathogen Effectors. The Plant cell *20*, 2009-2017. doi: 10.1105/tpc.108.060194.

Vetter, M.M., Kronholm, I., He, F., Haweker, H., Reymond, M., Bergelson, J., Robatzek, S., and de Meaux, J. (2012). Flagellin perception varies quantitatively in Arabidopsis thaliana and its relatives. Molecular biology and evolution *29*, 1655-1667. doi: 10.1093/molbev/mss011.

Vlot, A.C., Dempsey, D.A., and Klessig, D.F. (2009). Salicylic Acid, a multifaceted hormone to combat disease. Annual review of phytopathology *47*, 177-206. doi: 10.1146/annurev.phyto.050908.135202.

Vos, P., Simons, G., Jesse, T., Wijbrandi, J., Heinen, L., Hogers, R., Frijters, A., Groenendijk, J., Diergaarde, P., Reijans, M., *et al.* (1998). The tomato Mi-1 gene confers resistance to both root-knot nematodes and potato aphids. Nature biotechnology *16*, 1365-1369. doi: 10.1038/4350.

Wang, C., Zien, C.A., Afitlhile, M., Welti, R., Hildebrand, D.F., and Wang, X. (2000). Involvement of phospholipase D in wound-induced accumulation of jasmonic acid in arabidopsis. The Plant cell *12*, 2237-2246.

Wang, W., Wen, Y., Berkey, R., and Xiao, S. (2009). Specific targeting of the Arabidopsis resistance protein RPW8.2 to the interfacial membrane encasing the fungal Haustorium renders broad-spectrum resistance to powdery mildew. The Plant cell *21*, 2898-2913. doi: 10.1105/tpc.109.067587.

Wang, Y., Li, J., Hou, S., Wang, X., Li, Y., Ren, D., Chen, S., Tang, X., and Zhou, J.M. (2010). A Pseudomonas syringae ADP-ribosyltransferase inhibits Arabidopsis mitogen-activated protein kinase kinases. The Plant cell *22*, 2033-2044. doi: 10.1105/tpc.110.075697.

Weiberg, A., Wang, M., Lin, F.M., Zhao, H., Zhang, Z., Kaloshian, I., Huang, H.D., and Jin, H. (2013). Fungal small RNAs suppress plant immunity by hijacking host RNA interference pathways. Science (New York, NY) *342*, 118-123. doi: 10.1126/science.1239705.

Weiler, E.W., Kutchan, T.M., Gorba, T., Brodschelm, W., Niesel, U., and Bublitz, F. (1994). The Pseudomonas phytotoxin coronatine mimics octadecanoid signalling molecules of higher plants. FEBS letters *345*, 9-13.

Wirthmueller, L., Zhang, Y., Jones, J.D., and Parker, J.E. (2007). Nuclear accumulation of the Arabidopsis immune receptor RPS4 is necessary for triggering EDS1-dependent defense. Current biology : CB *17*, 2023-2029. doi: 10.1016/j.cub.2007.10.042.

Wroblewski, T., Caldwell, K.S., Piskurewicz, U., Cavanaugh, K.A., Xu, H., Kozik, A., Ochoa, O., McHale, L.K., Lahre, K., Jelenska, J., *et al.* (2009). Comparative large-scale analysis of interactions between several crop species and the effector repertoires from multiple pathovars of Pseudomonas and Ralstonia. Plant physiology *150*, 1733-1749. doi: 10.1104/pp.109.140251.

Wu, Y., Zhang, D., Chu, J.Y., Boyle, P., Wang, Y., Brindle, I.D., De Luca, V., and Despres, C. (2012). The Arabidopsis NPR1 protein is a receptor for the plant defense hormone salicylic acid. Cell reports *1*, 639-647. doi: 10.1016/j.celrep.2012.05.008.

Xiao, S., Calis, O., Patrick, E., Zhang, G., Charoenwattana, P., Muskett, P., Parker, J.E., and Turner, J.G. (2005). The atypical resistance gene, RPW8, recruits components of basal defence for powdery mildew resistance in Arabidopsis. The Plant journal : for cell and molecular biology *42*, 95-110. doi: 10.1111/j.1365-313X.2005.02356.x.

Xin, X.F., and He, S.Y. (2013). Pseudomonas syringae pv. tomato DC3000: a model pathogen for probing disease susceptibility and hormone signaling in plants. Annual review of phytopathology *51*, 473-498. doi: 10.1146/annurev-phyto-082712-102321.

Yamaguchi, Y., Pearce, G., and Ryan, C.A. (2006). The cell surface leucine-rich repeat receptor for AtPep1, an endogenous peptide elicitor in Arabidopsis, is functional in transgenic tobacco cells. Proceedings of the National Academy of Sciences of the United States of America *103*, 10104-10109. doi: 10.1073/pnas.0603729103.

Yang, X., Wang, W., Coleman, M., Orgil, U., Feng, J., Ma, X., Ferl, R., Turner, J.G., and Xiao, S. (2009). Arabidopsis 14-3-3 lambda is a positive regulator of RPW8-mediated disease resistance. The Plant journal : for cell and molecular biology *60*, 539-550. doi: 10.1111/j.1365-313X.2009.03978.x.

Yeats, T.H., and Rose, J.K. (2013). The formation and function of plant cuticles. Plant physiology *163*, 5-20. doi: 10.1104/pp.113.222737.

Yu, A., Lepere, G., Jay, F., Wang, J., Bapaume, L., Wang, Y., Abraham, A.L., Penterman, J., Fischer, R.L., Voinnet, O., *et al.* (2013). Dynamics and biological relevance of DNA demethylation in Arabidopsis antibacterial defense. Proceedings of the National Academy of Sciences of the United States of America *110*, 2389-2394. doi: 10.1073/pnas.1211757110.

Yun, B.W., Feechan, A., Yin, M., Saidi, N.B., Le Bihan, T., Yu, M., Moore, J.W., Kang, J.G., Kwon, E., Spoel, S.H., *et al.* (2011). S-nitrosylation of

NADPH oxidase regulates cell death in plant immunity. Nature *478*, 264-268. doi: 10.1038/nature10427.

Zhang, J., Li, W., Xiang, T., Liu, Z., Laluk, K., Ding, X., Zou, Y., Gao, M., Zhang, X., Chen, S., *et al.* (2010). Receptor-like cytoplasmic kinases integrate signaling from multiple plant immune receptors and are targeted by a Pseudomonas syringae effector. Cell host and microbe *7*, 290-301. doi: 10.1016/j.chom.2010.03.007.

Zhang, J., Shao, F., Li, Y., Cui, H., Chen, L., Li, H., Zou, Y., Long, C., Lan, L., Chai, J., *et al.* (2007). A Pseudomonas syringae effector inactivates MAPKs to suppress PAMP-induced immunity in plants. Cell host and microbe *1*, 175-185. doi: 10.1016/j.chom.2007.03.006.

Zhang, Y., Zhu, H., Zhang, Q., Li, M., Yan, M., Wang, R., Wang, L., Welti, R., Zhang, W., and Wang, X. (2009). Phospholipase dalpha1 and phosphatidic acid regulate NADPH oxidase activity and production of reactive oxygen species in ABA-mediated stomatal closure in Arabidopsis. The Plant cell *21*, 2357-2377. doi: 10.1105/tpc.108.062992.

Zhang, Z., Wu, Y., Gao, M., Zhang, J., Kong, Q., Liu, Y., Ba, H., Zhou, J., and Zhang, Y. (2012). Disruption of PAMP-induced MAP kinase cascade by a Pseudomonas syringae effector activates plant immunity mediated by the NB-LRR protein SUMM2. Cell host and microbe *11*, 253-263. doi: 10.1016/j.chom.2012.01.015.

Zhao, B.Y., Ardales, E., Brasset, E., Claflin, L.E., Leach, J.E., and Hulbert, S.H. (2004). The Rxo1/Rba1 locus of maize controls resistance reactions to pathogenic and non-host bacteria. TAG Theoretical and applied genetics Theoretische und angewandte Genetik *109*, 71-79. doi: 10.1007/s00122-004-1623-y.

Zhao, C., Nie, H., Shen, Q., Zhang, S., Lukowitz, W., and Tang, D. (2014). EDR1 physically interacts with MKK4/MKK5 and negatively regulates a MAP kinase cascade to modulate plant innate immunity. PLoS Genetics *10*, e1004389. doi: 10.1371/journal.pgen.1004389.

Zhu, Z., Xu, F., Zhang, Y., Cheng, Y.T., Wiermer, M., Li, X., and Zhang, Y. (2010). Arabidopsis resistance protein SNC1 activates immune responses through association with a transcriptional corepressor. Proceedings of the National Academy of Sciences of the United States of America *107*, 13960-13965. doi: 10.1073/pnas.1002828107.

Zipfel, C. (2009). Early molecular events in PAMP-triggered immunity. Current opinion in plant biology *12*, 414-420. doi: 10.1016/j.pbi.2009.06.003.

Zipfel, C., Kunze, G., Chinchilla, D., Caniard, A., Jones, J.D., Boller, T., and Felix, G. (2006). Perception of the bacterial PAMP EF-Tu by the receptor EFR restricts Agrobacterium-mediated transformation. Cell *125*, 749-760. doi: 10.1016/j.cell.2006.03.037.

Zipfel, C., Robatzek, S., Navarro, L., Oakeley, E.J., Jones, J.D., Felix, G., and Boller, T. (2004). Bacterial disease resistance in Arabidopsis through flagellin perception. Nature *428*, 764-767. doi: 10.1038/nature02485.

CHAPTER 4

Endofungal Bacteria Increase Fitness of their Host Fungi and Impact their Association with Crop Plants

Ibrahim Alabid[1], Stefanie P. Glaeser[2] and Karl-Heinz Kogel[1]*

[1]Institute of Phytopathology, IFZ Research Centre for Biosystems, Land Use and Nutrition, Justus Liebig University, D-35392 Giessen, Germany
[2]Institute of Applied Microbiology, IFZ Research Centre for Biosystems, Land Use and Nutrition, Justus Liebig University, D-35392 Giessen, Germany

*Karl-Heinz.Kogel@agrar.uni-giessen.de

DOI: https://doi.org/10.21775/9781912530007.04

Abstract:
Endofungal bacteria are bacterial symbionts of fungi that exist within fungal hyphae and spores. There is increasing evidence that these bacteria, alone or in combination with their fungal hosts play a critical role in tripartite symbioses with plants, where they may contribute to plant growth and disease resistance to microbial pathogens. As the frequency of bacteria in fungi is commonly very low, breakthroughs in technology such as molecular taxonomy and laser scanning microscopy were required to establish the functional contribution of these bacteria in complex symbioses. Yet, the overall biological significance of endofungal bacteria is largely unknown and further progress in understanding is hampered by a very few biological systems where endofungal bacteria have been described mechanistically. We review here the current knowledge on endobacteria (EB) and their role in different types of fungal symbioses with plants. We show that various attempts to cure fungal cells from endobacteria failed, further suggesting that they play a crucial role in the symbiosis. Moreover, isolation of some of the endobacteria from their fungal hosts allowed confirming their autonomous beneficial activity such as plant growth promotion and resistance-inducing activity. The review addresses the potential agricultural significance of endofungal bacteria and their role in supporting sustainable agriculture by promoting plant growth, improving plant resistance, and decreasing yield loss caused by many microbial pathogens.

Endobacteria in plant-colonizing fungi

Endofungal bacteria inhabit the cytoplasm of fungal cells (Figure 1). They commonly establish beneficial relationships (positive symbioses) with their plant-colonizing host fungi thereby forming tripartite interactions that comprise the bacterium, the fungus and the plant (Perotto and Bonfante, 1997; Bonfante and Anca, 2009; Desirò et al., 2014; Moebius et al., 2014; Erlacher et al., 2015; Glaeser et al., 2016; Salvioli et al., 2016). From the historical perspective, Mosse (1970) was the first to describe intracellular structures very similar to bacteria, called Bacteria-Like Organisms (BLOs) inside fungal hyphae (Figure 2). Since then, BLOs and bacteria were detected in glomeromycotan arbuscular mycorrhiza

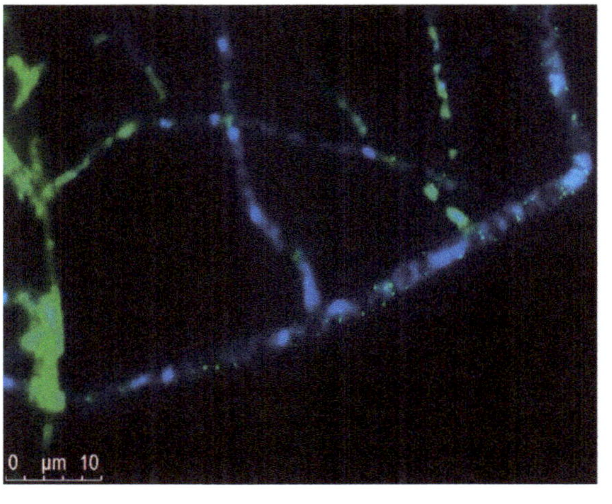

Figure 1. Detection of endofungal bacteria in fixed fungal mycelia of *Sebacina vermifera*-MAFF305838 by Fluorescence *in situ* hybridization (FISH) analysis using a universal Bacteria 16S rRNA targeting probe (green) and DAPI counter straining (blue). Detection of endobacteria in this strain was first described in Sharma et al. (2008).

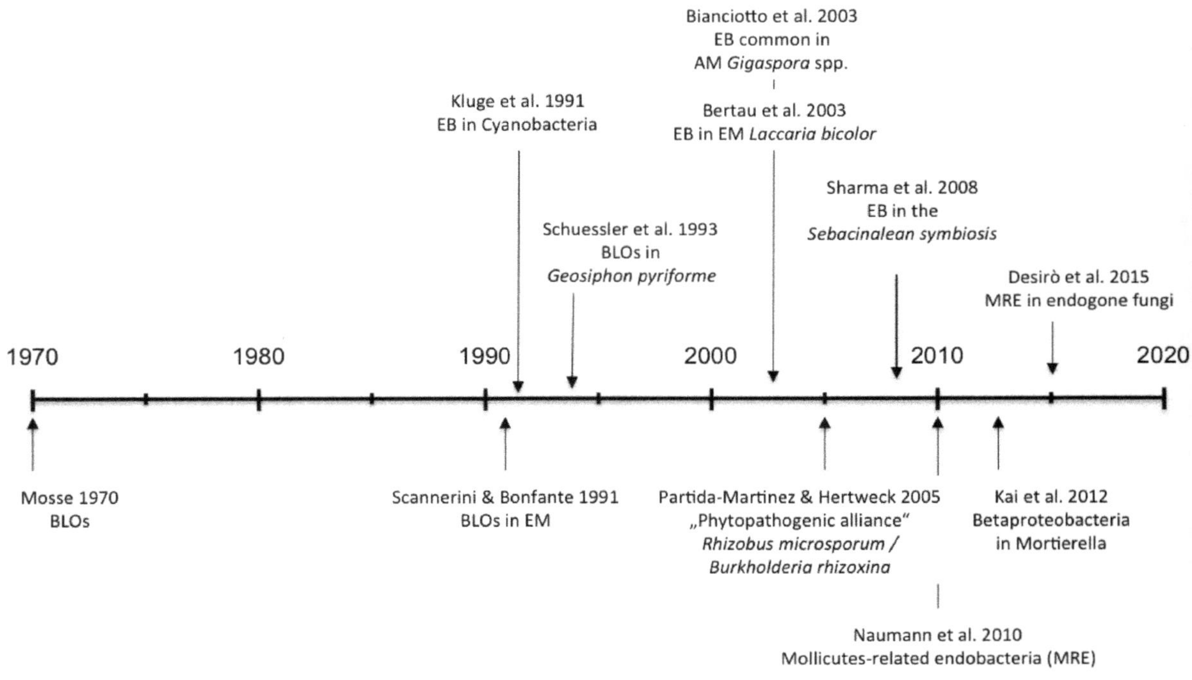

Figure 2. Time scale of the published research on endofungal bacteria.

(AM) symbioses (Scannerini and Bonfante, 1991; Bianciotto et al., 2003; Naumann et al., 2010), the related fungus *Geosiphon pyriforme* (Kluge, 1992; Schüßler et al., 1994), the ectomycorrhizal (EM) basidiomycete fungus *Laccaria bicolor* (Bertaux et al., 2003; 2005), the rice pathogenic fungus *Rhizopus microsporus* (Partida-Martinez and Hertweck, 2005), phylogenetically diverse foliar fungal endophytes (Hoffman and Arnold, 2010), and the plant symbiotic Endogone Mucoromycotina fungi (Desirò et al., 2015). Unlike insect endosymbionts, which are localized in specialized tissues (Moran et al., 2008), endofungal bacteria are found in fungal spores and both extra- and intraradical hyphae, the latter colonizing the plant tissues (Lumini et al., 2007). Endofungal bacteria also have been discovered in the Sebacinalean symbiosis, a mutualistic association of Basidiomycota fungi of the order Sebacinales with a broad spectrum of plants (Sharma et al., 2008). Although the role of these bacteria is not fully understood, increasing evidence gathered by studying the model Sebacinalean fungus *Piriformospora indica* suggests that they contribute to the fitness of their fungal hosts (Guo et al., submitted). Some reports illustrate that endofungal bacteria support the virulence of pathogenic fungi (Scherlach et al., 2006). In general, complex interactions including bacteria, fungi, and plants are poorly understood, and the global prevalence of these types of interactions is largely unknown. In this review we focus on the three most intensively studied biological models comprising endobacteria in beneficial and parasitic tripartite interactions (Figure 3).

Endofungal bacteria in the AM symbiosis

Two types of endofungal bacteria have been distinguished in AM fungi on the basis of their morphological features. The first type of endobacteria is coccoid-shape and present also in the cytoplasm of another basal group of fungi, the Endogone Mucoromycotina species. These bacteria represent a monophyletic clade of fungal mycelium-derived sequences which is phylogenetically placed in-between the Mollicutes and Firmicutes and termed Mollicutes-related endobacteria (MRE; Scannerini and Bonfante, 1991; Naumann et al., 2010; Desirò et al., 2015). In contrast to the cell-wall free Mollicutes, the MRE endobacteria detected in the cytoplasm of fungal cells contain a Gram-positive like cell wall.

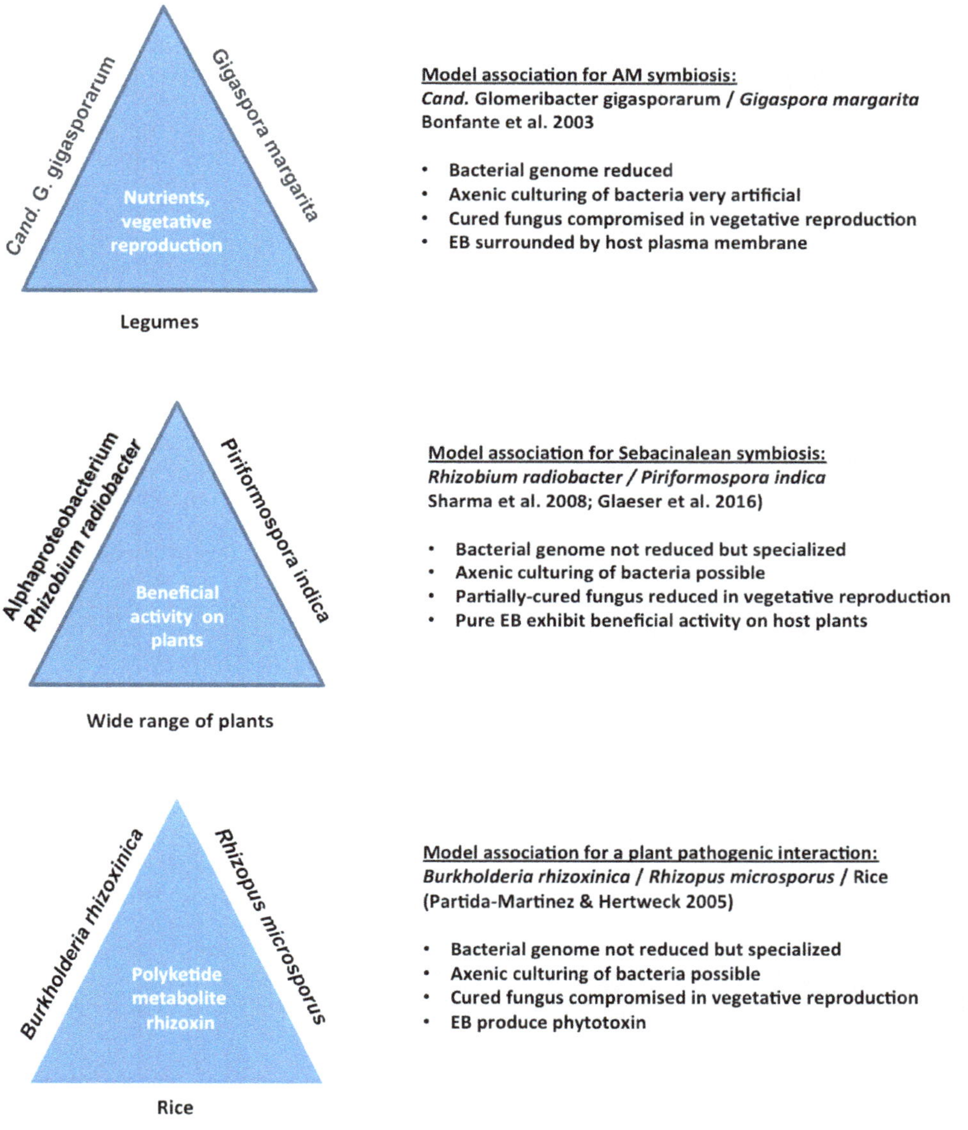

Figure 3. Comparative summary of the knowledge about the three most extensively studied tripartite symbioses of plant with their colonizing fungi and endobacteria.

The second type of endofungal bacteria are Gram-negative rod-shaped bacteria that are restricted to the AM-forming family *Gigasporaceae* (Bianciotto et al., 2004). Those intracellular bacteria were detected in five *Gigaspora* spp. through all stages of the fungal life cycle: spores, germtubes, extra- and intraradical hyphae. Based on the 16S rRNA gene sequence phylogeny, the endobacteria in *Gigaspora margarita* were initially identified as a member of the genus *Burkholderia* (Bianciotto et al., 1996). The *G. margarita* isolate BEG 34 contained 250,000 living bacterial cells in a single spore. The bacteria were subsequently classified based on genetic features as a novel bacterial taxon, *Candidatus* Glomeribacter gigasporarum (*Ca*Gg; Banciotto et al., 2003) next closest related to the genus *Burkholderia*. Isolated bacteria could not be grown in culture media, but kept alive for up to 4 weeks

(Jargeat et al., 2004). Sequencing of the genome of a homogeneous cell population derived from the *Gigaspora margarita* strain BEG 34 representing *Ca*Gg led to a 1.72 Mb assembly with 1,736 coding DNA sequences (CDS), the smallest genome known for a Betaproteobacterium (Ghignone et al., 2012). Such small genomes are typically found in endocellular bacteria living permanently in their host with an obligate symbiotic lifestyle, where gene erosion occured in association with metabolic dependence on the host (Moran et al., 2008; Castillo and Pawlowska, 2010). The genome assembly comprised one chromosome and three plasmids. Although the genome was rather small, the G+C content was high (54.8%), which is unusual for small genomes.

A phylogenetic multilocus gene analysis carried out on concatenated sequences of 21 protein-coding genes retrieved from 67 completed genomes belonging to a wide range of bacterial lineages showed that *Ca*Gg clusters within the family *Burkholderiaceae*. Members of this family are highly versatile microbes interacting with animals, humans, plants and fungi (Bontemps et al., 2010). Analysis of a data set restricted to *Betaproteobacteria* and based on 16S and 23S rRNA gene sequences placed *Ca*Gg as a sister group of a *Burkholderia* clade, which includes free-living and fungal-associated species showing that *Ca*Gg is an ancient member of the taxon sharing a common ancestor with the present-day *Burkholderiaceae* (Ghignone et al., 2012). While phylogenetic analyses placed *Ca*Gg in the *Burkholderiaceae*, metabolic pathway analyses clustered it with endosymbiotic bacteria of insects. This positioning among different bacterial classes let the authors claim that *Ca*Gg has undergone convergent evolution to adapt itself to an intracellular lifestyle. Genome annotation further revealed an unexpected genetic mosaic where typical genetic determinants of symbiotic, pathogenic and free living bacteria are integrated in a reduced genome.

Significantly, the bacterial cells are separated from the fungal cytoplasm by a membrane of fungal origin, suggesting that *Ca*Gg communicates with the fungus via transport and secretion systems that deliver bacterial molecules to the host. This is supported by the large number of genes present in its genome coding for secretion, including type II and type III secretion systems and synthesizes vitamin B12, antibiotics- and toxin-resistance molecules, which may contribute to the fungal host's ecological fitness (Ghignone et al., 2012). On the other hand, genome annotation also provided an insight into the molecular basis for *Ca*Gg obligate biotrophic status: The lack of some crucial metabolic pathways can explain the failure to grow *Ca*Gg as a free-living organism, as it has a metabolic dependence on the fungal host for both energy and nutrition. Thus, the bacterial genome provided clear evidence of the energy/nutrient flows of the tripartite interaction between the bacterium, the fungus and the plant. *Ca*Gg's limited capacity to synthesize amino acids, the presence of a large set of amino-acid transporters in the bacterial genome, and its location inside the protein-rich fungal vacuoles highlight that there is most likely a flow of nitrogen from the fungus to the bacterium. Consistent with the strong dependency, *Ca*Gg is at least in part but non-essential vertically transmitted through fungal vegetative sporulation, indicating that active bacterial proliferation occurs in the multinucleate mycelium of the fungus (Bianciotto et al., 2004). Consistent with this, the genome evolution in *Ca*Gg is non-degenerative and exemplifies a departure from the model of degenerative evolution in heritable endosymbionts such as mutualists of insects (Mondo et al., 2016).

While the occurrence of *Ca*Gg is limited to the *Gigasporaceae*, coccoid-shaped MRE endobacteria are widely distributed across different lineages of AM fungi. A fungus can harbor both types of endobacteria, with MRE population being more abundant, variable and prone to recombination (Desirò et al., 2014), suggesting that *Gigasporaceae* with their comparatively large spores, which are rich in reserves of glycogen, fats and proteins, can support the energetic cost of complex bacterial communities. MRE were also identified in the Endogone Mucoromycotina fungi, an ancient group of fungi capable of symbiotically interacting with plants. Interestingly, Mucoromycotina fungi along with Glomeromycota are considered as the unique ancestral symbionts of land plants (Desirò et al., 2015). MRE possess a homogenous cell wall-like envelope, which upon high pressure and freeze substitution is rather electron-transparent. 16S rRNA gene sequences of MRE isolated from AM fungi and Endogone Mucoromycotina fungi cluster together and form a separate clade. In both fungal groups, coccoid-shaped MRE are embedded in the fungal cytoplasm without any evidence of the fungal membrane which surrounds *Ca*Gg (Ghignone et al., 2012).

Endofungal bacteria impact the AM symbiosis

The mechanistic role of AM-associated endobacteria is still unclear, but their presence may be required for successful interaction of the fungi with their plant hosts. *In vitro* studies suggested that *Paenibacillus* sp. may stimulate AM's hyphal growth (Horii and Ishii 2006; Horii *et al.* 2008), the formation of new spores (Hildebrandt et al., 2002), and the suppression of pathogens (Budi et al., 1999; Horii et al., 2008). Though these bacteria were associated with the mycorrhizosphere, a true endofungal lifestyle has not been proven yet. Similarly, when surface-sterilized spores of *Glomus intraradices* Sy167 were germinated on agar plates, slime-forming bacteria, identified as *Paenibacillus validus*, frequently grew up. These bacteria supported the fungus completing its life cycle in the absence of plant roots (Hildebrandt et al., 2006).

Clear evidence for endofungal bacteria essentially contributing to beneficial AM symbioses is still missing because killing or separating bacteria from their AM host by antibiotic treatments or single spore proliferation is extremely difficult. The reason for this intimate association between endofungal bacteria and their hosts is not exactly known and may vary for single interaction systems. Yet, *G. margarita* BEG 34 provides a convincing example for a cured fungus: repeated passages through single-spore inocula caused dilution of the initial *Ca*Gg bacterial population eventually leading to cured AM spores. *Ca*Gg was not essential to the survival or reproduction of the fungus. However, spores had a distinct phenotype in terms of the cytoplasm organization, vacuole morphology, cell wall organization, lipid bodies and pigment granules. The absence of bacteria severely affected presymbiotic fungal growth such as hyphal elongation and branching. However, at least under laboratory conditions, cured *G. margarita* formed mycorrhizal associations and also sporulated in the colonized roots (Lumini et al., 2007). Recent work supported the view that *Ca*Gg increases the environmental fitness and bioenergetics potential of *G. margarita* by priming mitochondrial metabolic pathways (Salvioli et al., 2016). The endobacterium influenced fungal growth, metabolism, and calcium signaling by targeting mitochondrial activity, upregulation of the genes involved in respiration, ATP production and reactive oxygen species (ROS) detoxification (Vannini et al., 2016). The authors concluded that - at least for the *G. margarita* BEG 34 isolate - the absence of endofungal bacteria causes delays in the growth of germinating mycelium, possibly affecting its overall ecological fitness.

Endofungal bacteria in EM symbiosis

In contrast to AM symbiosis, a relatively small number of plants live in symbiosis with ectomycorrhizal (EM) fungi, mainly forest trees in temperate and boreal ecosystems. EM symbioses are formed by a large number of fungal species, mainly Basidiomycetes, but also Ascomycetes. In the early 1990th the concept of mycorrhization helper bacteria (MHB) emerged to describe bacteria that help to establish a mycorrhizal symbiosis (Garbaye, 1994; Frey-Klett et al., 2007). This concept comprised all types of mycorrhiza and free as well as endocellular bacteria that support a symbiosis and thus appears somewhat indistinct. We focus here mainly on reports suggesting that EM also is associated with true endofungal bacteria. A rather specific though not endocellular association was described for the bacterium *Streptomyces* strain AcH 505 that improves mycelial growth and EM formation between the fly agaric fungus *(Amanita muscaria)* and spruce (*Picea abies*). Auxofuran was identified as the predominant growth promoting substance which was most effective at a concentration of 15 µM. Co-cultivation of the bacteria and *A. muscaria* stimulated auxofuran production in the fungus. Another example is the contribution of *Pseudomonas fluorescens* BBc6R8, isolated from a *Laccaria bicolor* sporocarp consistently promoting *L. bicolor*-Douglas fir (*Pseudotsuga menziesii*) EM formation. The prevalence of strain BBc6R8 in the soil was significantly enhanced by the presence of *L. bicolor* S238N in either the presence or absence of Douglas-fir roots, while in contrast its survival was not supported by non-mycorrhizal roots, suggesting that the strain *P. fluorescens* BBc6R8 depends more on the fungus than on the plant roots (Frey-Klett et al., 2007).

Bacterial proliferations also have recurrently been observed in fermentor cultures of the *L. bicolor* strain S238N, suggesting cryptic bacteria associated with this fungus. Endofungal bacteria were detected by Fluorescence in situ Hybridization (FISH) in pure fungal subcultures. They were small in size (0.5 µm in diameter), rare, and heterogeneously distributed in the mycelium. 16S rRNA gene sequence analysis identified those endobacteria as *Paenibacillus* spp. (Bertaux et al., 2003). However, *Paenibacillus* spp. have not been recovered from colonized plant samples, although pure fungal cultures served as inoculum (Bertaux et al., 2005). Instead, samples taken from EM roots rather contained endofungal *Alphaproteobacteria*, which *vice versa* had not been detected in pure *L. bicolor* S238N cultures. Many *Alphaproteobacteria* also were detected outside the hyphae, in addition to bacteria belonging to other phyla, such as Actinomycetes and *Cytophaga-Flexibacter*. Thus, the authors speculated about an environmental origin of the endofungal *Alphaproteobacteria* (Bertaux et al., 2005).

Endofungal bacteria in the Sebacinalean symbiosis

The order Sebacinales is the most basal Basidiomycota group which contains fungi that undergo endophytic as well as mycorrhizal interactions with a broad spectrum of monocotyledonous and dicotyledonous plants found across all continents (Selosse et al., 2007; Weiss et al., 2011, Riess et al., 2014). The family *Serendipitaceae* comprises endophytes from the genus *Serendipita* such as *Piriformospora* (syn. *Serendipita*) *indica* that constitutes an excellent root endophyte model (Verma et al., 1998), especially due to available genome information, genetic tractability, and its broad host range that includes many important cereals and *Brassicaceae* such as the model plant *Arabidopsis thaliana* (for review Qiang et al., 2012). Unlike AM, *Serendipita* species are facultative biotrophic and thus can easily be cultured with synthetic medium in the absence of a plant (Deshmukh et al., 2006; Oelmüller et al., 2009). Endofungal bacteria identified in various *Serendipita* species belong to two genera of Gram-negative (*Rhizobium* and *Acinetobacter*) and two genera of Gram-positive (*Paenibacillus* and *Rhodococcus*) bacteria (Sharma et al., 2008; see figure 1). The most comprehensively studied example of a tripartite Sebacinalean symbiosis, including endofungal bacteria, is the association of *P. indica* with the Alphaproteobacterium *Rhizobium radiobacter* (*Rr*, syn. *Agrobacterium tumefaciens*, syn "*Agrobacterium fabrum*"; Figure 4). FISH using the *Rhizobium*-specific probe *Rhi-1247* confirmed the endocellular association of *Rr* with fungal spores and hyphae. Using quantitative PCR analysis, a ratio of 0.02 - 0.035 ng of bacterial DNA per 100 ng of *P. indica* DNA was determined (Sharma et al., 2008). This value is consistent with the low number of bacteria (2 - 20 per fungal cell) detected in *L. bicolor* (Bertaux et al., 2003; 2005). *R. radiobacter* was determined in the original strain of *P. indica*

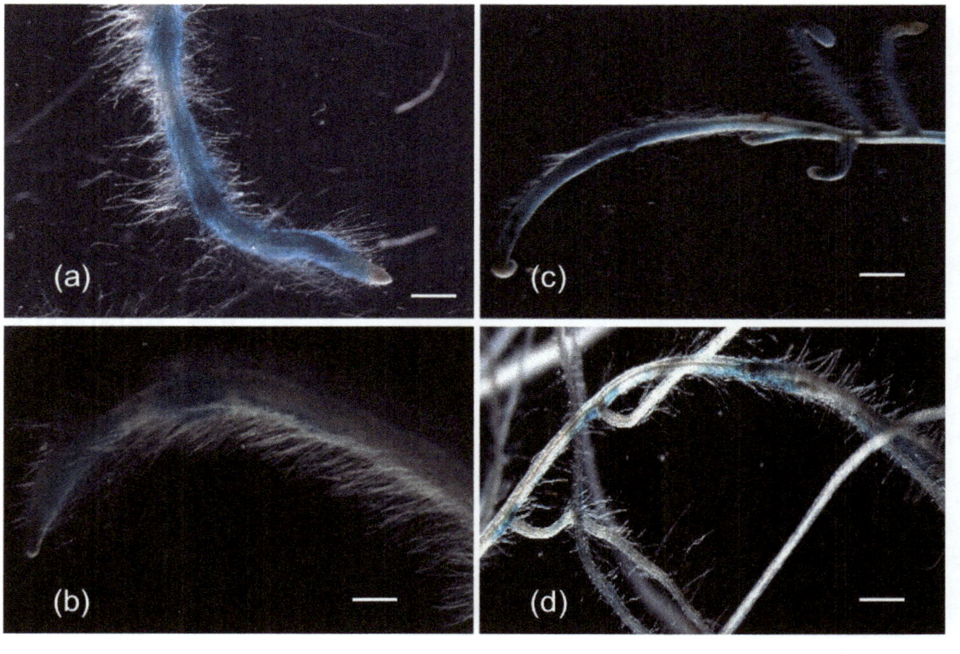

Figure 4. Colonization pattern upon inoculation of plant roots with the endofungal bacterium *R. radiobacter* F4 (*Rr*F4) that was isolated from a fungal *Piriformospora indica* culture. The endobacteria were labelled with the β-glucuronidase (GUS) reporter enzyme. Bacteria are visualized after X-Gluc treatment of roots at 7 (a,b) and 14 (c,d) days after dip inoculation. Staining of barley primary roots (a,b); lateral root protrusions and root hair zones of secondary roots of barley (c) and Arabidopsis (d). Scale bar = 2 mm.

DSM 11827 isolated from the Indian jar dessert and from three other *P. indica* cultures by the use of Denaturing gradient gel electrophoresis (DGGE) of 16S rRNA gene fragments amplified from fungal DNA extracts (Sharma et al., 2008). The strain *R. radiobacter* F4 (*Rr*F4) was isolated from *P. indica* DSM 11827 and propagated in axenic cultures with Luria-Bertani (LB) medium, showing that the bacterium is not entirely dependent on its fungal host. However, attempts to cure *P. indica* of *Rr*F4 have failed. Treatments with antibiotics such as spectinomycin and ciprofloxacin, which inhibit the growth of *Rr*F4 in pure cultures, merely transiently reduced the number of bacteria in the fungal mycelium as shown by the increase in the number of bacteria after prolonged cultivation in the absence of antibiotics. The fact that bacteria could not be killed inside the hyphae raised the speculation that *Rr*F4 enters an inactive state where it becomes insensitive to antibiotics. The protection against antibiotics may also be a survival strategy for bacteria in the rhizosphere which is commonly rich in antimicrobial compounds producing rhizobacteria. Nevertheless, a combination of protoplastation and antibiotics treatments reduced the abundance of endobacterial cells below the detection limit. Fungi regenerated from those protoplasts showed both reduced vegetative growth and chlamydospore formation suggesting a requirement of endobacteria for the fungal fitness (Guo et al., submitted). This situation is reminiscent of interactions of *Ca*Gg with G. *margarita* and *Burkholderia* spp. with *Rhizopus microspores*, where the absence of the respective bacterium also reduced fungal fitness (Lumini et al., 2007; Lackner et al., 2011).

The genome of *Rr*F4 is organized in a circular (2.8 Mb) and a linear chromosome (2.06 Mb), a tumor-inducing plasmid pTiF4 (0.21 Mb), and an accessory plasmid pAtF4 (0.54 Mb) and thus shows a high degree of similarity to the plant pathogenic *R. radiobacter* C58 (*Rr*C58; formerly: *Agrobacterium tumefaciens*; syn. "*A. fabrum"* C58; Goodner et al., 2001). The circular and linear chromosomes of *Rr*F4 had 100 and 80 singleton open reading frames (ORFs), respectively, that are not present in C58. Most of these ORFs were of unknown function and may be candidates for future studies to elucidate a potential role for the endofungal growth of *Rr*F4 and/ or fitness of its fungal partner *P. indica* (Glaeser et al., 2016). Differences such as the loss of the T-DNA in the tumor-inducing (pTi) plasmids can explain the loss of *Rr*F4's pathogenicity (Glaeser et al., 2016). In contrast to obligate endofungal bacteria such as the non-cultivable *Ca*Gg, the genome of *Rr*F4 is not reduced, which is consistent with the situation in the cultivable endobacterium *Burkholderia rhizoxinica* (Lackner et al., 2011). Thus, the data are consistent with the hypothesis that *Rr*F4 forms a facultative symbiosis with *P. indica*, where the bacterium is still able to live independently outside its host. Supportive for this hypothesis, curing *Rr*C58 from its pTi plasmid resulted in a non-pathogenic strain with weak plant growth-stimulating activity in maize seedlings in non-sterile soil (Walker et al., 2013).

Endofungal bacteria enhance plant growth

The benefits or costs that fungal endophytes and endophyte-bacterial complexes extend to their plant hosts are greatly associated with plant health and seed production. Thus, it is important to address the effects endofungal bacteria have on both, disease resistance and yield. Fungi and bacteria can interact synergistically to stimulate plant growth through a range of mechanisms that include improved nutrient acquisition and inhibition of fungal pathogens (Artursson et al., 2006). In general, the beneficial potential of bacteria closely associated with the plant root can be inferred from the group of plant growth-promoting rhizobacteria (PGPR) that are in contact with the root surface, or rhizoplane, and increase plant yield by diverse mechanisms such as improved mineral uptake, disease suppression, or phytohormone production (Weller, 1988; Kloepperet al., 1991; Lugtenberg et al., 1991; Broek and Vanderleyden, 1995; Défago and Keel, 1995). Even so they are not "endofungal", PGPR have been reported to benefit associated fungi during interaction with their plant host. Certain groups of bacteria have been shown to accumulate to a higher extent in the mycorrhizosphere compared with other groups (Artursson et al., 2006 and references therein). PGPR also interact with AM fungi and have stimulatory impact on the growth of these fungi suggesting that they have at least an additional indirect effect on plant growth promotion (Garbaye, 1994). For example, association of *Pseudomonas putida* with AM fungi result in increased growth of clover plants, suggesting that PGPR may have properties that support both mycorrhizal establishment and function (Meyer and Linderman, 1986). Phylogenetically diverse endofungal bacteria detected in hyphae of diverse foliar Ascomycota endophytes of trees produced the plant-growth promoting hormone indole-3-acetic acid (IAA). Most of these bacteria were members of the Proteobacteria (Hoffman and Arnold, 2010). For instance, *in vitro* production of IAA by a fungal endophyte determined as *Pestalotiopsis* spp. (Pezizomycotina) that was isolated from foliage of a coniferous host (*Platycladus orientalis*) was enhanced by the presence of *Luteibacter* sp. of the *Xanthomonadales*. IAA production by the endophyte-bacterial complex required L-tryptophan. The isolated and axenically cultured *Luteibacter* sp. did not produce IAA on a standard growth medium. However, culture filtrate from the endophyte-bacterium complex enhanced growth of tomato plants relative to filtrate from the endophyte alone. Given that the hormone produced by diverse plant-associated fungi can enhance plant growth and also suppresses, to a certain extent, plant defense such as the hypersensitive reaction (HR) and production of pathogenesis-related (PR) proteins, IAA production could be an important aspect of foliar endophyte-plant symbioses (Maor et al., 2004; Spaepen and Vanderleyden, 2011; Hoffman et al., 2013).

Inoculation of various plant species such as barley, wheat, and Arabidopsis with pure cultures of *Rhizobium radiobacter* F4 (*Rr*F4) promotes shoot and root growth, including lateral root branching (Sharma et al., 2008; Glaeser et al., 2016). These activities widely mimicked the effects *Rr*F4's host fungus *P. indica* has on plants. Yet, as the fungal host could not be completely cured from the endobacterium, it is still an unresolved question of whether beneficial plant biomass enhancement in the Sebacinalean symbiosis merely stems from *Rr*F4, the fungus *P. indica*, or the endophyte-bacterium complex (Guo et al., submitted).

In a forest nursery, the amount of phosphorus-solubilizing and siderophore-producing fluorescent pseudomonads was much higher in the Douglas-fir ectomycorrhizal fungus *L. bicolor* than in the surrounding root-free soil (Frey-Klett et al., 2005). The enhancement of the solubilization of rock phosphate occurred following formation of mixed biofilms between phosphate-solubilizing saprotrophic fungi and a *Bradyrhizobium elkanii* strain (Jayasinghearachchi and Seneviratne, 2005). The bacterial secondary metabolite responsible for plant growth promoting activity of *Amantia muscaria* through *Streptomyces* sp. AcH 505 is 5,6,7-trihydro-7-hydroxy-3-prolylbenzofuran-4-1, termed auxofuran, because of its auxin-reminiscent structure (Frey-Klett et al., 2007; Keller et al., 2006; Riedlinger et al., 2006). Upon treatment with auxofuran, expression of the *A. muscaria* gene acetoacetyl-CoA synthetase (*AmAacs*) was up-regulated, indicating activation of sterol biosynthesis (Riedlinger et al., 2006). Auxins and ethylene also have been implicated in inducing morphological changes in roots during mycorrhizal formation (Kaska et al., 1999), including the formation of lateral roots and dichotomous branching of short roots (Barker and Tagu, 2000).

Promotion of lateral root formation also is a commonly observed characteristic of MHB, essentially leading to an increase in potential contact points at which the plant and the ectomycorrhizal fungus can interact (Poole et al., 2001; Schrey et al., 2005). MHB can indirectly facilitate EM fungi's root colonization by inducing the release of signal molecules such as plant flavonoids (Frey-Klett et al., 2007) or suppressors of the plant's defence responses (Lehr et al., 2007). Moreover, root colonization by EM fungi may be facilitated be MHB's production of plant cell wall-digesting enzymes thereby enhancing penetration and spreading of the fungus within the root tissues (Mosse, 1962).

Acyl homoserine lactone production and quorum sensing of endofungal bacteria

Acyl homoserine lactones (AHLs) are well-known molecules produced by bacteria for their own communication termed quorum sensing (QS). More recently, AHLs were also implicated in beneficial activities bacteria have on plant yield and health (Schuhegger et al., 2006; Schikora et al., 2011; Schenk et al., 2014; Zarkani et al., 2013). Using a bacterial biosensor screening, AHL-producing endofungal bacteria were detected in antagonistic soil fungi (Kai et al., 2012). *In situ* detection by FISH analysis and electron microscopy confirmed the presence of endobacteria in the mycelium of the zygomycete fungus *Mortierella alpina* A-178 that produced the QS molecules. Release of these molecules by the fungus was subsequently proven by the detection of AHL from the supernatant of the liquid culture of the fungus. Amplification of 16S rRNA gene sequence fragments from fungal mycelium extracts indicated that a Beta-proteobacterium with 100% 16S rRNA gene sequence identity to *Castellaniella defragrans* (100 %) and a Gram-positive bacterium assigned to the genus *Cryobacterium* (99.8% 16S rRNA gene sequence identity) were present in the mycelium of *M. alpina* A-178. Antibiotic treatment enabled to cure the fungus, and the lack of AHL detection in the fungus confirmed that the AHL is produced by the endobacterium. Yet, how QS is involved in the interaction of fungus and bacterium needs to be elucidated by further studies. Similarly pure cultures of endofungal *Rr*F4, isolated from *P. indica*, also produce various classes of AHL that are known to have growth promoting and induced resistance activity in plants (unpublished data). Significantly, mutants of *Rr*F4 that were compromised in AHL accumulation, showed reduced growth promoting activity when colonizing wheat or Arabidopsis plants. These merely preliminary findings suggest a more rigorous analysis of the role of AHLs in tripartite interactions. Noteworthy, *B. rhizoxinica*, lacks genes related to known QS systems (Lackner et al., 2011b).

Complexes of closely associated bacteria and beneficial fungi enhance local and systemic resistance to microbial pathogens

Fungus-bacterium complexes also contribute to plant protection against leaf and root pathogens. Various studies have demonstrated *in vitro* antagonistic activity exerted by mycorrhiza-associated bacteria on microbial plant pathogens (Schelkle and Peterson, 1996; Becker et al., 1999; Maier et al., 2004; Frey-Klett et al., 2007). The EM-associated *Pseudomonas fluorescens* strain BBc6R8 inhibited the growth of various root-pathogenic fungi belonging to the genera *Rhizoctonia*, *Fusarium*, *Phytophthora* and *Heterobasidion* in the Douglas-fir - *L. bicolor* interaction (Frey-Klett et al., 2005). In addition, a strain of *Paenibacillus* spp. isolated from the rhizosphere of sorghum proved to be compatible with AM development, but inhibited soil-borne fungal pathogens (Budi et al., 1999). This bacterial strain produced small peptides that were harmless to the symbiotic fungi, but were responsible for the antagonistic effects to pathogens (Selim et al., 2005). Cruz and Ishii (2008) isolated two bacteria from surface sterilized spores of *Gigaspora margarita*. Both isolates that were closest related to *Janthinobacterium* and *Paenibacillus* spp. inhibited growth of the soil-borne plant pathogens *Fusarium oxysporum* f. sp. *lactucae*, *Rhizoctonia solani*, and *Pythium ultimum* (Cruz and Ishii, 2008).

Pure cultures of the endofungal bacterium *Rr*F4 induced systemic resistance against various microbial leaf pathogens such as the powdery mildew fungus *Blumeria graminis* f.sp. *hordei* and the bacterium *Xanthomonas translucens* pv. *translucens* (*Xtt*), when inoculated to barley or wheat roots, respectively. Similarly, *Rr*F4 also induced resistance in Arabidopsis to *Pseudomonas syringae* pv. *tomato* DC3000 (*Pst*) (Sharma et al., 2008; Glaeser et al., 2016). The underlying resistance mechanism in Arabidopsis depends on an operable jasmonate pathway, while mutants defective in genes governing SA accumulation and signaling where not compromised in *Rr*F4-mediated resistance (Glaeser et al., 2016).

Toxin-producing endofungal bacteria contribute to fungal virulence

The rice seedling blight fungus *Rhizopus microspores* (Mucoromycotina, Mucorales) severely affects rice fields in Asia that begins with an abnormal swelling of seedling roots and eventually results in death of the affected tissue or the complete plant. The fungus forms beneficial associations with the *Betaproteobacteria Burkholderia endofungorum* and *Burkholderia rhizoxinica* (Partida-Martinez et al., 2007b), which support the fungal host to parasitize plant tissue. *B. rhizoxinica* produces the phytotoxic polyketide metabolite rhizoxin (Partida-Martinez and Hertweck, 2005; Lackner et al., 2011) which blocks mitosis in eukaryotic cells, including plants, by binding to β-tubulin and thus serves the fungus as virulence factor (Sato *et al.* 1983). Due to an amino acid change in the tubulin protein, *Rhizopus* itself is tolerant to rhizoxin (Schmitt et al., 2008). Remarkably, rhizoxin had earlier been related to the fungal metabolism. However, treatment of the fungus with ciprofloxacin (40 μg ml^{-1}), an antibiotic active against bacteria but not against fungi, cured the fungus of the bacteria and concomitantly extirpated rhizoxin production (Partida-Martinez and Hertweck, 2005; Partida-Martinez et al., 2005). Consistent with this, *B. rhizoxinica* contains the gene cluster encoding rhizoxin biosynthesis (Partida-Martinez and Hertweck, 2007).

In contrast to the AM endofungal bacterium *Ca*Gg that resisted cultivation in cell-free medium and could only be investigated by using molecular methods, *B. rhizoxinica* could be isolated, grown in pure culture, and eventually reintroduced into a cured fungal host strain (Partida-Martinez and Hertweck, 2005; Scherlach et al., 2006; Partida-Martinez et al., 2007; Moebius et al., 2014). The bacterium is vertically transmitted through asexual reproduction units (vegetative spores) of its fungal host, a process that was discovered through re-infection of fungal mycelia using laser-mediated microinjection or co-cultivation with green fluorescent protein (GFP)-producing bacterial strains (Partida-Martinez et al., 2007a). Interestingly, in the absence of the endofungal bacteria, *R. microsporus* is compromised in asexual reproduction as evidences by a lack of mature sporangia and spores in mycelium of strains lacking bacteria. Formation of sporangia was restored upon reintroduction of bacteria, showing that reproduction of the fungal host is strictly dependent on the intact association (Partida-Martinez et al., 2007a). Symbiont-dependent sporulation is a hallmark of close mutualistic relationships (Moran, 2006) and an elegant way to prevent formation of symbiont-free spores, thus securing the persistence of the symbiosis. Further experiment showed that the bacterial rhizoxin is not required for the bacterial infection process nor for fungal sporulation: Both the *B. rhizoxinica* wild-type strain and a mutant deficient in rhizoxin biosynthesis supported fungal sporulation in culture showing that other bacteria factors are necessary for the symbiosis. Supporting this notion, a Type III secretion system (T3SS) was discovered in *B. rhizoxinica* that controls a range of interactions between the bacterium and its fungal host, including invasion, intracellular transport of bacteria to hyphal tips and sporulation (Lackner et al., 2011). The entire gene cluster spans about 22,000 bp and contains 23 open reading frames (ORFs). In terms of primary sequence conservation and gene order, the gene cluster is similar to the hypersensitive response protein (hrp) locus coding for a T3SS that promotes virulence of the plant pathogen *Ralstonia solanacearum* (Cunnac et al., 2004). *B. rhizoxinica* T3SS$^-$ mutants defective in two T3SS genes, *DsctC* and *DsctT*, did not show growth defects or morphological phenotypes compared with the wild type when grown in culture. However, these mutants were strongly compromised in their fitness as only limited zones of infection were visible. As in cured fungal cultures, the fungal host infected with these mutants was affected in intracellular survival and failed to elicit formation of mature sporangia in an infection assay (Lackner et al., 2011).

Isolation of fungal endobacteria

Only few endofungal bacteria have been isolated from their host fungi. Cruz et al. (2008) described the isolation of endofungal bacteria from *Gigaspora margarita*. After bacteria were determined by PCR from surface sterilized spores, they were isolated by osmosis from fungal protoplasts. The protoplasts were generated from spores by using two enzymes, a lysing enzyme and catalase. Two oval cells forming bacteria were isolated by this cultivation approach, a Gram-negative bacterium closest related to *Janthinobacterium lividum* and a Gram-positive bacterium closest related to *Paracoccus polymyxa*. Both isolates showed antagonistic properties against the pathogenic fungi *Rosellinia necatrix*, *Pythium ultimum*, *Fusarium oxysporum* and *Rhizoctonia solani*.

The endobacterium *Rr*F4 was isolated in a similar manner from crushed mycelium of *P. indica* (Sharma et al., 2008), and the two endosymbiotic *Burkholderia* species, *B. rhizoxinica* and *B. endofungorum,* were isolated from the supernatant of disrupted mycelium of two strains of *R. microspores*. Whether the free living growth state of endobacteria occurs in nature as a part of the bacterial live cycle is unclear. Glaeser et al. (2016) speculated that the endobacterium of *P. indica* may be released by root colonizing fungi to penetrate into internal root tissue to enhance the biological activity. A clear proof by *in situ* detection is still missing.

Bacterial invasion into fungal cells

Although endofungal bacteria have been known for four decades, the mechanism by which they enter fungal cells is unresolved. A recent study showed that a type II secretion system (T2SS) of the endofungal bacteria *B. rhizoxinica* is required for the formation of the endosymbiosis (Moebius et al., 2014), based on genome mining for potential symbiosis factors and functional analyses. Comparative proteome analyses show that the bacterium releases chitinolytic enzymes (chitinase, chitosanase) and chitin-binding proteins. Genes encoding chitinase and chitosanase are highly expressed during the infection. The authors suggested that secreting these enzymes and presumably further effector proteins via a T2SS help to locally soften the fungal cell wall allowing bacterial entry and preventing the disintegration of fungal hyphae without permanent damage (Moebius et al., 2014). Consistent with this, a chitinase loss-of-function mutant lost its ability to enter fungal hyphae.

Conclusion

Elucidating the biology of tripartite associations between plants, higher fungi, and endofungal bacteria has shown that endofungal bacteria have various beneficial activities that support growth and development of their fungal hosts. The association of *Rhizopus microsporus* with *B. rhizoxinica* is an outstanding model, because the fungus can be cured from the endofungal bacteria and isolated bacteria can be grown in pure cultures, two prerequisites to fully elucidate the role of bacteria in fungi. That is why the role of endofungal bacteria in endo-, ecto-, and Sebacinalean symbioses is still hampered as the fungi either cannot by cured or endobacteria cannot be cultured outside the fungal mycelium. Accordingly, our understanding of these tripartite symbioses is far from being complete. New strategies are needed to remove the bacteria from the fungal cytoplasm to enable comparisons of fungal effects on plants in the presence and absence of the bacterial symbionts. Despite these setbacks, it seems a more common phenomenon that root endophytic fungi host endobacteria that enhance the fungal fitness. In the cases the endofungal bacteria could be isolated and grown in pure culture they show similar biological activities than the fungal hosts itself. This makes it feasible that the fungus gains its full plant-colonizing activity from its intricate interaction with endobacteria.

Beneficial effect of pure endofungal bacteria also make them potentially interesting as application as biologicals that could further minimize the use of chemical fertilizers and increase the resistance of crop plants to a wider range of pathogens and pest, where the mechanism greatly relays on their direct antagonistic activity and their resistance-inducing potential. The present knowledge on endobacteria already shows the requirement towards unraveling a complex network of interphylum interactions, which is expected to have a previously unrecognized ecological impact.

References:

Artursson, V., Finlay, R.D., and Jansson, J.K. (2006). Interactions between arbuscular mycorrhizal fungi and bacteria and their potential for stimulating plant growth. Environ. Microbiol. *8*, 1-10.

Barker, S., and Tagu, D. (2000). The roles of auxins and cytokinins in mycorrhizal symbioses. Journal of Plant Growth Regulation *19*, 144-154.

Becker, D.M, Bagley, S.T., and Podila, G.K. (1999). Effects of mycorrhiza-associated streptomycetes on growth of *Laccaria bicolor*, *Cenococcum geophilum*, and *Armillaria* species and on gene expression in *Laccaria bicolor*. Mycologia *91*, 33-40.

Bertaux, J., Schmid, M., Hutzler, P., Hartmann, A., Garbaye, J., and Frey-Klett, P. (2005). Occurrence and distribution of endobacteria in the plant-associated mycelium of the ectomycorrhizal fungus *Laccaria bicolor* S238N. Environ. Microbiol. *7*, 1786-1795.

Bertaux, J., Schmid, M., Prévost-Bourre, N.C., Churin, J.L., Hartmann, A., Garbaye, J. and et al. (2003). *In situ* identification of intracellular bacteria related to *Paenibacillus* spp. in the mycelium of the ectomycorrhizal fungus *Laccaria bicolor* S238N. App. Environ. Microbiol. *69*, 4243-4248.

Bianciotto, V., Bandi, C., Minerdi, D., Sironi, M., Tichy, H.V., and Bonfante, P. (1996). An obligately endosymbiotic mycorrhizal fungus itself harbors obligately intracellular bacteria. Appl. Environ. Microbiol. *62*, 3005-3010.

Bianciotto, V., Genre, A., Jargeat, P., Lumini, E., Becard, G., and Bonfante, P. (2004). Vertical transmission of endobacteria in the arbuscular mycorrhizal fungus *Gigaspora margarita* through generation of vegetative spores. Appl. Environ. Microbiol. *70*, 3600-3608.

Bianciotto, V., Lumini, E., Bonfante, P., and Vandamme, P. (2003). *Candidatus Glomeribacter gigasporarum gen. nov., sp. nov.*an endosymbiont of arbuscular mycorrhizal fungi. Int. J. Syst. Evol. Microbiol. *53*, 121-124.

Bonfante P., and Anca, I.A. (2009). Plants, mycorrhizal fungi, and bacteria: a network of interactions. Annu. Rev. Microbiol. *63*, 363-383.

Bontemps, C., Elliott, G.N., Simon, M.F., Dos Reis, F.B.D., Gross, E., Lawton, R.C. et al. (2010). *Burkholderia* species are ancient symbionts of legumes. Mol. Ecol. *19*, 44-52.

Broek, A.V. and Vanderleyden, J. (1995). The role of bacterial motility, chemotaxis and attachment in bacteria plant interactions, MPMI. *8*, 800-810.

Budi, S.W., van Tuinen, D., Martinotti, G., and Gianinazzi, S. (1999). Isolation from the *Sorghum bicolor* mycorrhizosphere of a bacterium compatible with arbuscular mycorrhiza development and antagonistic towards soilborne fungal pathogens. Applied and Environmental Microbiology *65*, 5148-5150.

Castillo D.M., and Pawlowska, T.E. (2010). Molecular Evolution in Bacterial Endosymbionts of Fungi. Mol. Biol. Evol. *27*, 622-636.

Cruz, A. F., Horii, S., Ochiai, S., Yasuda, A., and Ishii, T. (2008). Isolation and analysis of bacteria associated with spores of *Gigaspora margarita*. J. Appl. Microbiol. *104*, 1711-1717.

Cunnac, S., Occhialini, A., Barberis. P., Boucher, C., and Genin, S. (2004). Inventory and functional analysis of the large Hrp regulon in *Ralstonia solanacearum*: identification of novel effector proteins translocated to plant host cells through the type III secretion system. Molecular Microbiology *53*, 115-128.

Défago, G., and Keel, C. (1995). Pseudomonads as biocontrol agents of diseases caused by soil-borne pathogens. pp. 137-148 in Hokkanen HMT, Lynch JM (eds), Biological Control: Benefits and Risks. Plant and Microbial Biotechnology Research Series 4. Cambridge University Press. ISNB 9780521544054.

Deshmukh, S.D., Hückelhoven, R., Schäfer, P., Imani, J., Sharma, M., Weiss, M. and et al. (2006). The root endophytic fungus Piriformospora indica requires host cell death for proliferation during mutualistic symbiosis with barley. Proc. Natl. Acad. Sci. USA. *103*, 18450-18457.

Desirò, A., Faccio, A., Kaech, A., Bidartondo, M.I, and Bonfante, P. (2015). Endogone, one of the oldest plant-associated fungi, host unique Mollicutes-related endobacteria. New Phytologist *205*, 1464-1472.

Desirò, A., Salvioli, A., Ngonkeu, E.L, Mondo, S.J., Epis, S., Faccio, A., Kaech, A., Pawlowska, T.E., and Bonfante, P. (2014). Detection of a novel intracellular microbiome hosted in arbuscular mycorrhizal fungi. ISME. J. *8*, 257-270.

Erlacher, A., Cernava, T., Cardinale, M., Soh, J., Sensen, C.W., Grube, M., and Berg, G. (2015). *Rhizobiales* as functional and endosymbiontic members in the lichen symbiosis of *Lobaria pulmonaria* L. Front. Microbiol. *6*, 53.

Frey-Klett, P., Chavatte, M., Clausse, M.L., Courrier, S., Le Roux, C., Raaijmakers, J., Martinotti, M.G., Pierrat, J.C., and Garbaye, J. (2005). Ectomycorrhizal symbiosis affects functional diversity of rhizosphere fluorescent pseudomonads. New Phytol. *165*, 317-328.

Frey-Klett, P., Garbaye, J., and Tarkka, M. (2007). Tansley review: the mycorrhiza helper bacteria revisited. New Phytologist *176*, 22-36.

Garbaye, J. (1994). Helper bacteria: a new dimension to the mycorrhizal symbiosis. New Phytol. *128*, 197-209.

Ghignone, S., Salvioli, A., Anca, I., Lumini, E., Ortu, G., Petiti, L., and et al. (2012). The genome of the obligate endobacterium of an AM fungus reveals an interphylum network of nutritional interactions. ISME. J. *6*, 136-145.

Glaeser, S.P, Imani, J, Alabid, I., Guo, H.J., Kumar, N., Kämpfer, P., Hardt, M., Blom, J., Rothballer, M., Hartmann, A., and Kogel, K.H. (2016). Non-pathogenic *Rhizobium radiobacter* F4 deploys plant beneficial activity independent of its host *Piriformospora indica*. ISME J. *10*, 871-884.

Goodner, B., Hinkle, G., Gattung, S., Miller, N., and Blanchard, M. (2001). Genome sequence of the plant pathogen and biotechnology agent *Agrobacterium tumefaciens* C58. Science *294*, 2323-2328.

Guo, H.J, Glaeser, S.P, Alabid, I., Imani, J., Haghighi, H., Kämpfer, P., and Kogel, K.H. (2016). The incidence of endofungal bacterium *Rhizobium radiobacter* (syn. *Agrobacterium tumefaciens*) increases in its fungal host *Piriformospora indica* at the onset of the tripartite Sebacinalean symbiosis with higher plants. Front. Microbiol, Submitted.

Hildebrandt, U., Janetta, K., and Bothe, H. (2002). Towards growth of arbuscular mycorrhizal fungi independent of a plant host. Appl. Env. Microbiol. *68*, 1919-1924.

Hildebrandt, U., Ouziad, F., Marner, F.J, and Bothe, H. (2006). The bacterium *Paenibacillus validus* stimulates growth of the arbuscular mycorrhizal fungus *Glomus intraradices* up to the formation of fertile spores. FEMS Microbiol. Lett. *254*, 258-67.

Hoffman M.T., Gunatilaka M.K., Wijeratne, K., Gunatilaka, L., and Arnold A.E. (2013). Endohyphal bacterium enhances production of indole-3-acetic acid by a foliar fungal endophyte. PLOS ONE *8*, e73132.

Hoffman, M.T., and, Arnold, A.E. (2010). Diverse bacteria inhabit living hyphae of phylogenetically diverse fungal endophytes. Appl. Environ. Microbiol. *76*, 4063-4075.

Horii, S., and Ishii, T. (2006). Identification and function of *Gigaspora margarita* growth-promoting microorganisms. Symbiosis *41*, 135-141.

Horii, S., Matsumura, A., Cruz, A., Ishii, T. (2008). Effect of arbuscular mycorrhizal fungi and their helper microorganisms on root colonization and growth of trifoliate orange seedlings. Proc. Int. Soc. Citriculture *1*, 567-571.

Jargeat, P., Cosseau, C., Ola'h, B., Jauneau, A., Bonfante, P., Batut, J., and Becard, G. (2004). Isolation, free-living capacities, and genome structure of '*Candidatus* Glomeribacter gigasporarum', the endocellular bacterium of the mycorrhizal fungus *Gigaspora margarita*. J. Bacteriol. *186*, 6876-6884.

Jayasinghearachchi, H.S., and Seneviratne, G. (2005). Fungal solubilization of rock phosphate is enhanced by forming fungal-rhizobial biofilms. Soil Biology and Biochemistry *38*, 405-408.

Kai, K., Furuyabu, K., Tani, A., and Hayashi, H. (2012). Production of the quorum-sensing molecules N-acylhomoserine lactones by endobacteria associated with *Mortierella alpina* A-178. Chembiochem *13*, 1776-1784.

Kaska, D.D., Myllyla, R. and Cooper, J.B. (1999). Auxin transport inhibitors act through ethylene to regulate dichotomous branching of lateral root meristems in pine. New Phytologist *142*, 49-58.

Keller, S., Schneider, K., and Sussmuth, R.D. (2006). Structure elucidation of auxofuran, a metabolite involved in stimulating growth of fly agaric, produced by the mycorrhiza helper bacterium *Streptomyces* AcH 505. Journal of Antibiotics (Tokyo) *59*, 801-803.

Kloepper, J.W., Rodrguez-kbana, R., Mcinroy, J.A., and Collins, D.J. (1991). Analysis of populations and physiological characterization of microorganisms in rhizospheres of plants with antagonistic properties to phytopathogenic nematodes. Plant Soil *136*, 95-102.

Kluge, M., Mollenhauer, D., Mollenhauer R., and Kape R. (1992). *Geosyphon pyriforme*, an endosymbiotic consortium of a fungus and a cyanobacterium (Nostoc), fixes nitrogen. Bot. Acta. *105*, 343-344.

Kluge, M., Mollenhauer, D., and Mollenhauer, R. (1991). Photosynthetic carbon assimilation in Geosiphon pyriforme (Kützing) F. v. Wettstein, an endosymbiotic association of fungus and cyanobacterium. Planta *185*, 311-315.

Lackner, G., Moebius, N., and Hertweck, C. (2011). Endofungal bacterium controls its host by an hrp type III secretion system. ISME. J. *5*, 252-261.

Lackner, G., Moebius, N., Partida-Martinez, L.P., Boland, S., and Hertweck, C.(2011b). Evolution of an endofungal Lifestyle: Deductions from the *Burkholderia rhizoxinica* Genome. BMC Genomics *12*, 210.

Lehr, N.A., Schrey, S.D., Bauer, R., Hampp, R., and Tarkka, M.T. (2007). Suppression of plant defense response by a mycorrhiza helper bacterium. New Phytol. *174*, 892-903.

Lugtenberg, B.J.J., de Weger, L.A., and Bennett, J.W. (1991). Microbial stimulation of plant growth and protection from disease. Current Opinion in Biotechnology *2*, 457-464.

Lumini, E., Bianciotto, V., Jargeat, P., Novero, M., Salvioli, A., Faccio, A., et al. (2007). Presymbiotic growth and sporal morphology are affected in the arbuscular mycorrhizal fungus *Gigaspora margarita* cured of its endobacteria. Cellu. Microbiol. *9*, 1716-1729.

Maier, A., Riedlinger, J., Fiedler, H.P., and Hampp, R. (2004). Actinomycetales bacteria from a spruce stand: characterization and effects on growth of root symbiotic and plant parasitic soil fungi in dual culture. Mycological Progress *3*, 129-136.

Maor, R., Haskin, S., Levi-Kedmi, H., Sharon, A. (2004). *In planta* production of indole-3-acetic acid by *Colletotrichum gloeosporioides* f sp *aeschynomene*. Appl. Environ. Microbiol. *70*, 1852-1854.

Meyer, R.J., and Linderman, R.G. (1986). Response of subterranean clover to dual inoculation with vesicular-arbuscular mycorrhizal fungi and plant growth-promoting rhizobacterium, *Pseudomonas putida*. Soil Biol. Biochem. *18*, 185-190.

Moebius, N., Üzüm, Z., Dijksterhuis, J., Lackner, G., and Hertweck, C. (2014). Active invasion of bacteria into living fungal cells. Elife 3, e03007.

Mondo, S.J, Salvioli, A., Bonfante, P., Morton, J.B, and Pawlowska, T.E. (2016). Nondegenerative Evolution in Ancient Heritable Bacterial Endosymbionts of Fungi. Mol. Biol Evol. *33*, 2216-2231.

Moran, N.A., McCutcheon, J.P., and Nakabachi, A. (2008) .Genomics and evolution of heritable bacterial symbionts. Annu. Rev. Genet. *42*, 165-190.

Moran, N.A. (2006). Symbiosis, Curr. Biol. *16*, R866-R871.

Mosse, B. (1962). The establishment of vesicular-arbuscular mycorrhiza under aseptic conditions. Journal of General Microbiology *27*, 509-520.

Mosse, B. (1970). Honey-coloured, sessile Endogone spores: II. Changes in fine structure during spore development. Arch. Microbiol. *74*, 129-145.

Naumann, M., Schüssler, A., and Bonfante, P. (2010). The obligate endobacteria of arbuscular mycorrhizal fungi are ancient heritable components related to the Mollicutes. ISME. J. *4*, 862- 871.

Oelmüller, R., Sherameti, I., Tripathi, S., and Varma, A. (2009). *Piriformospora indica*, a cultivable root endophyte with multiple biotechnological applications. Symbiosis *49*, 1-17.

Partida-Martinez, L.P., and Hertweck, C. (2005). Pathogenic fungus harbours endosymbiotic bacteria for toxin production. Nature *437*, 884-888.

Partida-Martinez, L.P., and Hertweck, C. (2007a). A Gene Cluster Encoding Rhizoxin Biosynthesis in "*Burkholderia rhizoxina*", the Bacterial Endosymbiont of the Fungus *Rhizopus microsporus*. ChemBioChem *8*, 41-45.

Partida-Martinez, L.P., Groth, I., Schmitt, I., Richter, W., Roth, M., and Hertweck, C. (2007b). *Burkholderia rhizoxinica* and *Burkholderia endofungorum*, bacterial endosymbionts of the rice pathogenic fungus *Rhizopus microsporus*. Int. J. Syst. Evol. Microbiol. *57*, 2583-2590.

Partida-Martinez, L.P., Monajembashi, S., Greulich, K.O., and Hertweck, C. (2007a). Endosymbiont-dependent host reproduction maintains bacterial-fungal mutualism. Curr. Biol. *17*, 773-777.

Perotto, S., and Bonfante, P. (1997). Bacterial associations with mycorrhizal fungi: close and distant friends in the rhizosphere. Trends in Microbiology *5*, 496-503.

Poole, E.J., Bending, G.D., Whipps, J.M., and Read, D.J. (2001). Bacteria associated with *Pinus sylvestris-Lactarius rufus* ectomycorrhizas and their effects on mycorrhiza formation *in vitro*. New Phytologist *151*, 743-751.

Qiang, X., Weiss, M., Kogel, K.H., and Schäfer, P. (2012). *Piriformospora indica*-a mutualistic basidiomycete with an exceptionally large plant host range. Molecular Plant Pathology *13*, 508-518.

Riedlinger, J., Schrey, S.D., Tarkka, M.T., Hampp, R., Kapur, M., and Fiedler, H.P. (2006). Auxofuran, a novel metabolite that stimulates the growth of fly agaric, is produced by the mycorrhiza helper bacterium *Streptomyces* strain AcH 505. App. Environ. Microbiol. *72*, 3550- 3557.

Riess, K., Oberwinkler, F., Bauer, R., and Garnica, S. (2014). Communities of endophytic sebacinales associated with roots of herbaceous plants in agricultural and grassland ecosystems are dominated by *Serendipita herbamans* sp. nov. PLoS One *9*, e94676.

Salvioli, A., Ghignone, S., Novero, M., Navazio, L., Venice, F., Bagnaresi, P., and Bonfante. P. (2016). Symbiosis with an endobacterium increases the fitness of a mycorrhizal fungus, raising its bioenergetics potential. ISME J. *10*, 130-144.

Sato, Z., Noda, T., Matsuda, I., Iwasaki, S., Kobayashi, H., and et al. (1983). Studies on rhizoxin, a phytotoxin produced by *Rhizopus chinensis* causing rice seedling blight. Annu. Phytopathol. Soc. Japan *49*, 128.

Scannerini, S., and Bonfante, P. (1991). Bacteria and bacteria like objects in endomycorrhizal fungi (Glomaceae) in: Margulis, L., Fester, R. (eds) Symbiosis as a source of evolutionary innovation: speciation and morphogenesis. MIT press, Cambridge, Mass, pp273-287.

Schelkle, M., and Peterson, R.L. (1996). Suppression of common root pathogens by helper bacteria and ectomycorrhizal fungi *in vitro*. Mycorrhiza *6*, 481-485.

Schenk, S.T., Hernández-Reyes, C., Samans, B., Stein, E., Neumann, C., Schikora, M., Reichelt, M., Mithöfer, A., Becker, A., Kogel, K.H., and Schikora, A. (2014). N-Acyl-homoserine lactone primes plants for cell wall reinforcement and induces resistance to bacterial pathogens via the salicylic acid/oxylipin pathway. Plant Cell *26*, 2708-2723.

Scherlach, K., Partida-Martinez, L.P., Dahse, H.M., and Hertweck, C. (2006). Antimitotic rhizoxin derivatives from a cultured bacterial endosymbiont of the rice pathogenic fungus *Rhizopus microsporus*. J. Am. Chem. Soc. *128*, 11529-11536.

Schikora, A., Schenk, S.T, Stein, E., Molitor, A., Zuccaro, A., and Kogel, K.H. (2011). N-acyl-homoserine lactone confers resistance towards biotrophic and hemibiotrophic pathogens via altered activation of AtMPK6. Plant Physiol. *157*, 1407-1418.

Schmitt, I., Partida-Martinez, L.P., Winkler, R., Voigt, K., Einax, E., and *et al.* (2008). Evolution of host resistance in a toxin-producing bacterial-fungal alliance. ISME. J. *2*, 632-641.

Schrey, S.D., Schellhammer, M., Ecke, M., Hampp, R., and Tarkka, M.T. (2005). Mycorrhiza helper bacterium *Streptomyces* AcH 505 induces differential gene expression in the ectomycorrhizal fungus *Amanita muscaria*. New Phytologist *168*, 205-216.

Schuhegger, R., Ihring, A., Gantner, S., Bahnweg, G., Knappe, C., Vogg, G., Hutzler, P., Schmid, M., Van Breusegem, F., Eberl, L., and et al. (2006). Induction of systemic resistance in tomato by N-acyl-L-homoserine lactone-producing rhizosphere bacteria. Plant Cell Environ. *29*, 909-918.

Schüßler, A., Mollenhauer, D., Schnepf, E., and Kluge, M. (1994). *Geosiphon pyriforme*, an endosymbiotic association of fungus and cyanobacteria: the spore structure resembles that of arbuscular mycorrhizal (AM) fungi. Botanica Acta. *107*, 36-45.

Selim, S., Negrel, J., Govaerts, C., Gianinazzi, S., and van Tuinen, D. (2005). Isolation and partial characterization of antagonistic peptides produced by *Paenibacillus* sp. strain B2 isolated from the *Sorghum* mycorrhizosphere. Applied and Environmental Microbiology *71*, 6501-6507.

Selosse, M.A., Setaro, S., Glatard, F., Richard, F., Urcelay, C., and Weiß, M. (2007). *Sebacinales* are common mycorrhizal associates of Ericaceae. New Phytol. *174*, 864-878.

Sharma, M., Schmid, M., Rothballer, M., Hause, G., Zuccaro, A., Imani, J., Kämpfer, P., Domann, E., Schafer, P., Hartmann, A., and Kogel, K.H. (2008). Detection and identification of bacteria intimately associated with fungi of the order Sebacinales. Cellular Microbiology *10*, 2235-2246.

Spaepen, S, and Vanderleyden, J. (2011). Auxin and Plant-Microbe Interactions. Cold Spring Harb Perspect Biol. *3*, a001438.

Vannini, C., Carpentieri, A., Salvioli, A., Novero, M., Marsoni, M., Testa, L., Concetta de Pinto, M., Amoresano, A., Ortolani, F., Bracale, M., and Bonfante, P. (2016). An interdomain network: the endobacterium of a mycorrhizal fungus promotes antioxidative responses in both fungal and plant hosts. New Phytol. *211*, 265-275.

Verma, S., Varma, A., Rexer, K.H., Hassel, A., Kost, G., and Sarabhoy, A. (1998). *Piriformospora indica*, gen. et sp. nov., a new root-colonizing fungus. Mycologia *90*, 896-903.

Walker, V., Bruto, M., Bellvert, F., Bally, R., Muller, D., Prigent-Combaret, C., Moënne- Loccoz, Y., and Comte, G. (2013). Unexpected phyto-stimulatory behavior for *Escherichia coli* and *Agrobacterium tumefaciens* model strains. Mol. Plant Microbe Interact. *26*, 495-502.

Weiss, M., Selosse, M.A., Rexer, K.H., Urban, A., and Oberwinkler, F. (2011). Sebacinales everywhere: previously overlooked ubiquitous fungal endophytes. PLoS One *108*, 1003-1010.

Weller, D.M. (1988). Biological control of soilborne plant pathogens in the rhizosphere with bacteria. Annual Review of Phytopathology *26*, 379-407.

Zarkani, A.A., Stein, E., Rohrich, C.R., Schikora, M., Evguenieva-Hackenberg, E., Degenkolb, T., and et al. (2013). Homoserine lactones influence the reaction of plants to rhizobia. Int. J. Mol. Sci. *14*, 17122-17146.

CHAPTER 5

Plant-Nematode Interactions Assisted by Microbes in the Rhizosphere

Olivera Topalović[1]* and Holger Heuer[1]

[1]Julius Kühn-Institut, Messeweg 11-12, 38104 Braunschweig, Germany

*olivera.topalovic@julius-kuehn.de

DOI: https://doi.org/10.21775/9781912530007.05

Abstract

Plant health is strongly influenced by the interactions between parasites/pathogens and beneficial microorganisms. In this chapter we will summarize the up-to date knowledge on soil suppressiveness as a biological tool against phytonematodes and explore the nature of monoculture versus crop rotation in this regard. Since nematodes are successfully antagonized by different microbiological agents, we highlighted this phenomenon with respect to the most important antagonists, and a nature of these interactions. The focus is on the hyperparasitic microbes of phytonematodes such as *Pasteuria* sp. and egg parasites. Furthermore, we comprised the studies on the defence system expressions in plants triggered by nematode-associated microbes. The attachment of bacteria and fungi to phytonematodes and putative effects of the attachment on the induced systemic resistance in plants are discussed. Finally, our chapter is rounded up with the importance of incorporating the knowledge on plant-nematode-microbe interactions in the integrated pest management.

Introduction

Considering that every living macroorganism can be regarded as a holobiont as it is inevitably interconnected with its corresponding microbial cohabitants (Bordenstein and Theis, 2015; Vandenkoornhuyse et al., 2015), and transferring this phenomenon on the tripartite plant-nematode-microbiome system, we can assume that the destiny of plants attacked by plant-parasitic nematodes (PPN) much depends on which microbes are enriched endophytically and in the rhizosphere before and during parasitism events. PPN are considered one of the major pests of agricultural plants and it has been estimated that they cause yield losses up to $80 billion (Handoo, 1998). The majority of PPN belongs to the order Tylenchida, with the endoparasitic root-knot nematodes (RKN), *Meloidogyne* spp., cyst nematodes (CN), *Heterodera* spp. and *Globodera* spp., and root-lesion nematodes (RLN), *Pratylenchus* spp., being the most devastating phytonematodes (Nicol et al., 2011). The RKN and CN are sedentary endoparasites with infective second-stage juveniles (J2) which move through soil and infect roots of host plants. After reaching suitable root cells, they become sedentary, and start producing feeding sites, syncytia (CN) or giant cells (RKN). This results in nematode development into females that protrude egg masses inside or outside the root galls (RKN), or the eggs are encumbered in encysted female bodies (CN). The genus *Pratylenchus* includes migratory endoparasitic nematodes with all life stages (females, males, juveniles) being motile and infective.

Infective stages of PPN encounter a vast number of soil microbiota before entering plants. This allows the attachment of diverse microbes to their cuticle and surface coat. What nematodes are probably not aware of is that when allowing the attachment and agree on carrying the microbes to the plants, they may cause self-deleterious effects, i.e. they can be antagonized by transmitted microbiota. In the following lines, we will review the up-to-date knowledge on the importance and ability of specifically attached soilborne microbes to suppress invasion and reproduction of PPN in plants. Our intention is to confront the attainments, merits and demerits of a plant mediated versus direct antagonism of PPN, with respect to different nematode life strategies and antagonists involved. The molecular aspects of a cross-communication between the plant, nematodes and nematode-attached microbes will be discussed, and the benefits of nematode-attached microbes to the plant health will be summarized.

Nematode-suppressive soils

Two types of soil suppressiveness have been described in the literature. One is general (non-specific) soil suppressiveness which characterizes the ability of any soil to suppress many pathogens to a certain extent and is related to microbial activity (Davies and Spiegel, 2011). Specific soil suppressiveness is directed towards a particular pathogen. It is exerted by a complete absence of pathogen establishment or, in the worst-case scenario it can cause a severe disease and then draw back with a continued growing of the host plant (Baker and Cook, 1974). Both types of soil suppressiveness are removed by sterilization, indicating the biological nature of soil suppressiveness, but the mechanisms and which microbial species are involved is not sufficiently explored. The previous research was focused on direct antagonism of soil microbes towards soilborne phytopathogens, while plant-mediated mechanisms and plant-associated microbes have not been considered sufficient to explain and manage soil suppressiveness. The follow-up examples support the evidence that suppressive soils can be used as a powerful biocontrol measure against PPN. In an early study by Crump and Kerry (1987) a population of *Heterodera schachtii* failed to increase in suppressive soil of a 2-year monoculture of sugar beet and increased only in the third year. Treatment of this soil with the fungicide Captafol resulted in a greater nematode multiplication and a low fungal parasitism of females, suggesting that fungi played a role in this soil suppressiveness. The soil was enriched with the nematophagous fungi *Pochonia chlamydosporia*, *Fusarium* spp., and *Cylindrocarpon destructans*, which was isolated from infected females. These fungal species are well known parasites of CN (Jorgenson, 1970; Kerry, 1988). When Westphal and Becker (1999) had followed reproduction of *H. schachtii* on Swiss chard (*Beta vulgaris* L.) in a monoculture California field amended with *H. schachtii*-infested soil, the nematode population started declining after several years and remained very low in the following period. The soil suppressiveness was removed by biofumigation or a biocide-treatment. Similarly, a biological soil suppressiveness was responsible for a decreased reproduction of *H. schachtii* on Swiss chard (Westphal et al., 2011). The soil fumigation with iodide resulted in a higher nematode multiplication. In a rare example of soil suppressiveness against an ectoparasitic PPN the population density of *Criconemella xenoplax* dropped significantly in a suppressive soil between 1988-1991 (Kluepfel et al., 1993). Although edaphic factors were nearly identical for both suppressive and non-suppressive sites, it was unclear whether these two sites were indeed comparable, and nematode population densities were not reduced below the damage threshold in the fields. Taking into an account the feeding strategy of ectoparasitic nematodes, which do not enter the roots but feed only by piercing the epidermal root cells with their stylets, it should be noted that the soil microbes attached to the cuticle of these nematodes do not get into a direct contact with plants. This means that soil suppressiveness against ectoparasitic PPN might be inefficient because they are less prone to plant-mediated mechanisms of suppression.

On many occasions, it was proposed that the general indicators of the abundance and activity of soil microbiota could give an indication whether microbial factors are responsible for soil suppressiveness. For instance, a role of the blanket of sugarcane residues and mill mud amendments covering the surface of a sugarcane field was tested against the endoparasitic nematode *Pratylenchus zeae* (Stirling et al., 2011). Although more than 90% of the sugarcane root biomass occupied the depth just beneath the trash blanket, the nematode density at this depth was very low. One would expect that a high total carbon and the readily oxidisable carbon fraction measured within the upper soil profile when the soil surface mulched could have positively influenced soil microbial activity and reduced nematode population density at this depth (Stone et al., 2004; Stirling et al., 2005; Stirling et al., 2011). A similar observation was noticed after the density of *Pratylenchus thornei* had been depressed in upper soil layers of the wheat fields (Stirling, 2011).

The most common way to prove the biological nature of soil suppressiveness includes the comparison of nematode performance in sterilized and non-sterilized soil. In assessing the effects of gamma irradiation on soil properties, decreased levels of $NO3^-$ and an increase of $NH4^+$ were recorded due to elimination of denitrifying bacteria (Buchan et al., 2012). The absence of a microbial community also resulted in a high abundance of labile carbon content due to its insufficient utilization. This leads to the conclusion that the soil microbiota has a profound effect on plants by manipulating soil mineral components, and soil sterilization acts both against deleterious and beneficial microflora. For instance, a negative correlation was found between a shoot growth of white clover and soil gamma irradiation when studying soil suppressiveness against *Meloidogyne hapla* (Bell et al., 2016). Thus, in parallel with its increased attractiveness, soil suppressiveness still needs to be considered a sensitive topic. Further studies on the cooperation between different soil factors would arm us with a better knowledge on the successful manipulation of specific soil suppressiveness against PPN.

The key mechanisms underlying biological soil suppressiveness involve a direct antagonism of PPN by soil microbiota, and induced systemic resistance in plants (Sikora et al., 2007). To better understand how plants can benefit from the antagonistic interactions between microorganisms and PPN, following lines will review some bacterial and fungal groups that are found to suppress PPN, and the processes taking place beneath these interactions.

Egg parasites of phytonematodes

An extensive literature can be accessed regarding different bacterial and fungal facultative and obligate parasites of PPN (Stirling and Mankau, 1978; Nigh et al., 1980; Kerry, 1988; Meyer et al., 1990; Jaffee et al., 1992; Dijksterhius et al., 1994; Meyer and Wergin, 1998; Hidalgo-Diaz et al., 2000; Lopez-Llorca et al., 2002; Verdejo-Lucas et al., 2002; Olatinwo et al., 2006; Singh et al., 2007; Liu et al., 2009; Castillo et al., 2010; Davies and Spiegel, 2011; Kiewnick et al., 2011; Moosavi and Zare, 2012; Manzanilla-Lopez et al., 2013; Stirling, 2014; Hussain et al., 2017). As for molecular and biochemical aspects of nematode-parasite interaction, the most extensively studied obligate biotroph of PPN is the gram-positive bacterium *Pasteuria* sp. It forms endospores that attach to the nematode's cuticle while it is moving through soil. When the nematodes enter the plant, endospores penetrate the nematode body wall and reach the pseudocelom where they start to germinate and complete their life cycle. Three different species of *Pasteuria* have been recognized to antagonize PPN, *Pasteuria penetrans* on *Meloidogyne* spp., *Pasteuria thornei* on *Pratylenchus* spp. and *Pasteuria nishizawe* on *Heterodera* spp. and *Globodera* spp. (Chen and Dickson, 1998). The mechanism by which nematodes are colonized by this bacterium has been well studied (Chen and Dickson, 1998; Davies and Curtis, 2011; Stirling, 2014). The bacterial spores probably attach to glycoproteins of the surface coat of PPN which contain carbohydrate residues that specifically bind lectins of the bacterial surface (Spiegel et al., 1995). The recognition between epitopes present on the bacterial surface and on the surface coat of nematodes is an initial and essential step in the endospore attachment (Davies and Curtis, 2011), and the specificity of the *Pasteuria*-nematode association is very high and varies even between different individuals of the same species (Davies et al., 1994; Davies and Williamson, 2006; Davies et al., 2008). Although evidence is gathering on a successful control of PPN by high densities of *Pasteuria* spp. (Bird and Brisbane, 1988; Schmidt et al., 2010; Stirling, 2014), the variation in specificity of the attachment in combination with costly mass production of this obligate biotrophic microorganism *in vitro* imposes limitations for use of this hyperparasite as a broad-spectrum biocontrol agent against nematodes. Furthermore, the density of *Pasteuria* endospores is ironically dependent on the density of nematodes (prey) (Stirling, 2014). If the nematode population in soil is low, not many of them will encounter bacterial spores and further endospore propagation will be negligible. Conversely, when the majority of nematodes is being infected with bacteria, they will not only be subjected to its own decline, but will no longer represent the source of an increased endospore inoculum. A pronounced host specificity of this hyperparasitic bacterial group could also appear problematic in cases where mixed nematode populations cause disease in a certain plant crop, resulting in a selective advantage of populations being not a host for *Pasteuria*. So, although *Pasteuria* spores were responsible for the decline of a *Meloidogyne arenaria* race 1 population in a peanut crop, the yield losses were still apparent due to the presence of *Meloidogyne javanica* that could not host the same *Pasteuria* strain and resisted hyperparasitism (Cetintas and Dickson, 2004). Therefore, one needs to consider adjusting many factors before expecting a decline of PPN populations by *Pasteuria* sp. in different crops.

Many fungal isolates were also found to directly antagonize PPN. The roughest subdivision of nematophagous fungi involves: the nematode-trapping fungi which use specialized morphological structures, traps, to hook free-living nematodes, the endoparasitic fungi which infect nematodes by using adhesive spores, and the egg- and cyst-parasitic fungi (Barron, 1977). Several records witness PPN suppression by nematode-trapping fungi (Kerry, 1988; Kumar and Singh, 2006; Singh et al., 2007), and physiological, biochemical and molecular background of nematode trapping by these fungi have been described (Dijksterhius et al., 1994; Davies and Spiegel, 2011). A successful example is a significant decrease of the disease incidence and nematode fecundity on rice plants when the nematophagous fungi *Arthrobotrys dactyloides* and *Dactylaria brochopaga* were applied to soil infested with J2 of *Meloidogyne graminicola* (Singh et al., 2007). However, these fungi can only trap nematode migratory stages and species in soil, and the period of trapping activity might not be synchronized with the migration of *Meloidogyne* J2 to the roots, resulting in inefficient nematode control (Kerry, 1988). With respect to this, a much better alternative involves nematode control with obligate parasites of eggs and sedentary stages (females) in plants. The fungus *Pochonia chlamydosporia* and the oomycete *Dactylella oviparasitica* have been the most successful in colonizing and cutting down RKN and CN (Stirling and Mankau, 1978; Kerry, 1988; Verdejo-Lucas et al., 2002; Olatinwo et al., 2006). The examples of reduced nematode multiplication caused by these organisms and mechanisms on nematode parasitism are reviewed in books of Davies and Spiegel (2011) and Stirling (2014).

The application of facultative saprophytic fungi, *Trichoderma* sp., has also had a positive result in reducing RKN (Sharon et al., 2001), but the mode of suppression seems to be plant-mediated (Mukherjee et al., 2012; Martínez-Medina et al., 2013; Martínez-Medina et al., 2016; de Medeiros et al., 2017). The recognition between the fungus and the nematode is also carbohydrate-lectins dependent, as mentioned in case of *Pasteuria* sp. (Sharon et al., 2009). Sharon et al. (2007) have shown that the gelatinous matrix, in which the eggs of *Meloidogyne* are embedded, plays a key role in the agglutination of *Trichoderma* (except for *T. harzianum*) conidia in Ca^{2+}-dependent manner, and eggs separated from it are hardly infected by this fungus. This is inconsistent with the study by Orion et al. (2001), where eggs separated from the gelatinous matrix were heavily infected by soil microbiota, suggesting a protective role of the gelatinous matrix. However, we should not forget that *Trichoderma* sp. is a facultative saprophyte and may feed on the gelatinous matrix rather than parasitizing nematodes. Being a facultative saprophyte, which survives in the absence of nematodes but is still able to trigger plant responses against them, may have an advantage compared to obligate parasites of phytonematodes. Although obligate parasites have shown to be successful in suppressing PPN in greenhouse experiments, their mass production *in vitro* is the major problem when applying them as biocontrol agents. Their ability to outcompete the indigenous soil bacteria when introduced to the soil and the inoculum amount for an effective control, also need to be considered before their application (Eilenberg et al., 2001; Inceoglu et al., 2012).

The role of rhizobacteria in suppression of phytonematodes

The soil type and plant genotype determine which soil microbiota will be recruited by the plant and colonize the roots (Bulgarelli et al., 2012; Haney et al., 2015). Although the correspondence between soil-type-specific and root-endophytic bacteria indicate that most endophytes originate from soil and can be found in the rhizosphere (Bulgarelli et al., 2012), in this section we will discuss those bacteria that are antagonistic against PPN and isolated from plant rhizospheres. Nematodes penetrate plant root areas with a continuous excretion of root exudates. Bais et al. (2006) have recently reviewed the role of root exudates in interactions between plant, microbes and nematodes in the rhizosphere. These interactions can be fruitful for phytonematodes, when e.g. root volatiles attract cyst nematodes to the roots (Farnier et al., 2012), or deleterious, instanced by a negative effect of lauric acid from *Chrysantemum coronarium* L. on chemotaxis and infection ability of root-knot nematodes (Dong et al., 2014). As it is postulated that plants have evolved the ability to cultivate specific beneficial microbiomes within their rhizosphere (Haney et al., 2015), the follow-up examples show the ability of rhizomicrobes to act against PPN. Kloepper et al. (1992) studied the suppression of the phytonematodes *Heterodera glycines* and *Meloidogyne incognita* by isolated rhizobacteria from soybean (*Glycine max* L.) and several non-host plants of these nematodes, including velvet bean (*Muncuna deeringiana* L.), castor bean (*Ricinus communis* L.), sword bean (*Cannavalia ensiformis* L.) and Abruzzi rye (*Secale cereale* L.). With both nematodes, bacterial isolates from antagonistic plants reduced disease incidence 4-6 times compared to those from soybean, suggesting a contribution of these rhizobacteria to the non-host status of the plants they originated from. In another study, Gram-negative rhizobacteria were responsible for the reduction of the early infection of sugar beet in the greenhouse by *H. schachtii* (Ostendorp and Sikora, 1989). However, this effect was weakly expressed in field experiments for most bacterial isolates, especially in the second year after application. This was attributed to a high competition between the indigenous microflora and introduced isolates which could not well establish in the rhizosphere. As reported by Scher et al. (1988) the mean root colonization by different strains of *Pseudomonas putida*, *Pseudomonas fluorescens* and *Serratia* sp. was 1.0-1.5 log units lower per gram of root in a field compared to a greenhouse experiment. Furthermore, although two fluorescent pseudomonads, strains SS3 and 3K, failed to colonize maize roots in field soil, they reached population densities of >1x10^7 cfu/g root in autoclaved soil. The latter discussion requires consideration of many factors before designing experiments on nematode suppression by beneficial microbes, especially equalizing the greenhouse with real field conditions (Kloepper and Beauchamp, 1992; Eilenberg et al., 2001). Beside the ability to directly antagonize PPN, rhizomicrobiota can also enhance the systemic resistance of plants through mediation of jasmonic acid- and ethylene-pathways (van Loon et al., 1998), and some studies in this regard are gathered in the section on induced systemic resistance.

The role of plant endophytes in suppression of phytonematodes

Knowing that plant endophytes persist inside the plant itself, either through their entire life cycle or during a certain life stage, we cannot use this term to make a distinction between plant deleterious and plant mutualistic inhabitants. However, in this section we will elaborate the plant bacteria and fungi that live internally, within plant tissues (endorhiza), but for which it has been reported to suppress PPN. Sikora and Pocasangre (2006) have emphasized the importance of the so called pathozone, or the zone in soil used for initial root growth which is exposed to an early infection by nematodes. This zone can extend by 25-50 cm in soil from the plant base, while the application of pesticides mostly covers the upper 25 cm of soil. Considering that nematodes can move up to 50 cm in soil over a seven-day period, it means that newly formed roots remain unprotected from the nematode attack. Thus, a seek for effective mutualistic plant endophytes which are able to antagonize PPN would overcome the aforementioned problem (Sikora and Pocasangre, 2006).

Plant endophytic bacteria

Maybe the only advantage that plants can get from the PPN is the introduction of beneficial endophytes through the wounds that nematodes make by piercing the root cells. As mentioned in the previous section, it has been suggested that most of the endophytic bacteria cannot be marked as obligate endophytes because they also inhabit the rhizosphere and phylloplane (Kloepper and Beauchamp, 1992; Sikora et al., 2007; Bulgarelli et al., 2012). The attachment of beneficial endophytes to the cuticle of motile nematode stages in soil can presumably also lead to their transmission inside the roots. Although a knowledge gap still exists in this area, nematode interactions with plant endophytic bacteria have been observed by several authors (Hallmann et al., 1997; Hallmann et al., 1999; Hallmann et al., 2001; Siddiqui and Shaukat, 2003a; Sikora et al., 2007; Schouten, 2016). There are several known mechanisms by which endophytic bacteria can suppress plant infection caused by PPN, including competition for niche, direct antagonism by toxic compounds and induced systemic resistance (ISR) (Hallmann et al., 2001). However, a high density of bacterial antagonists (opportunistic bacteria) not always results in nematode suppression, and a direct toxicity might be compromised by a low expression of toxic compounds *in planta* (Sikora et al., 2007). This means that the induced systemic resistance might be the most important way of nematode control by endophytic bacteria (see section 6).

Plant endophytic fungi

Recently, Schouten (2016) has collected a vast literature on the involvement of endophytic fungi on PPN suppression, and covered the most important mechanisms that contribute to it. Recalling the concept of a holobiont (Bordenstein and Theis, 2015), mycorrhizal fungi represent an inextricable part of almost every plant system. Their role in suppression of PPN has been extensively studied and reviewed (Sitaramaiah and Sikora, 1982; Saleh and Sikora, 1984; Cooper and Grandison, 1987; Hol and Cook, 2005; de la Pena et al., 2006; Sikora et al., 2007; Sikora et al., 2008; Veresoglou and Rillig, 2011; Banuelos et al., 2014; Schouteden et al., 2015; Schouten, 2016). Hol and Cook (2005) have listed all recorded interactions between endoparasitic or ectoparasitic PPN and arbuscular mycorrhizal fungi (AMF), noting that the association of plants with AMF decreased the damage caused by sedentary endoparasites (RKN, CN) but increased the damage caused by ectoparasitic nematodes. They considered the possibility that ectoparasitic nematodes damage the AMF hyphae while browsing and, thus, evade the plant defense reactions. Numbers of migratory endoparasitic nematodes (*Pratylenchus, Radopholus*) were slightly increased on AMF-infected plants compared to AMF-free plants, but the damage was low in the reviewed studies. The mode of action of mycorrhizal fungi against PPN can be exhibited through a direct effect, by competition for nutrients and space, or indirect effect, by increasing plant tolerance, mediating ISR in plants, altering rhizosphere interactions by altered root exudations, or all of these combined (Schouten, 2016). The altered root exudation of AMF plants can directly affect nematode hatching, motility, chemotaxis, and host location (see (Stirling, 2014; Schouteden et al., 2015)). The ability of AMF to act against phytonematodes by mediating plant responses is discussed in the next section.

The association between phytonematodes and grass endophytes has also been reviewed (Cook and Lewis, 2001; Schouten, 2016). The most prominent genus is *Neothypodium*. Although it was reported to suppress some PPN, including *Meloidogyne* and *Pratylenchus* (Ball et al., 1997; Elmi et al., 2000; Jia et al., 2013), it has been understood that in some other cases this was not true (Nyczepir and Meyer, 2010), possibly due to plant genotype differences and different fungal isolates (Cook and Lewis, 2001). As studied by Meyer et al. (2013) the compounds from tall fescue root exudates and extracts reduced *M. incognita* vitality and fecundity in laboratory and greenhouse tests. However, a clear mode of action of these fungi against nematodes has not been understood yet and knowledge gaps on responsible fungal isolates prevents their incorporation in commercial plant cultivars (Stirling, 2014). There are certain indications which place some antagonistic *Fusarium* sp. strains into a group of plant-mediators when reducing reproduction of *Meloidogyne* spp. (Martinuz et al., 2013; Le et al., 2016). A more elaborative discussion on mediating plant responses by endophytic bacterial and fungal antagonists against PPN is presented in the next section.

Do microbiomes associated with infective stages of PPN trigger plant defenses?

From a myriad of microorganisms present in soil, only a small subset manages to bind to the nematode's cuticle. Despite several studies which indicate how the actual microbial attachment to PPN occurs, the questions have remained unanswered on what drives this high binding specificity and determines which microbes will be attached and which not. Further research on the nematode/ microbial selectivity in the attachment and the influence of plants on these interactions would better elucidate possibilities in manipulation of this binding for plant health purposes.

As summarized in previous sections, most studies on soil suppressiveness have tested an antagonistic potential of responsible microorganisms through their direct effect on PPN, including parasitism or production of toxins and enzymes (Mazzola, 2004; Westphal, 2005). However, evidence is mounting that plant-mediated mechanisms play a pivotal role in nematode suppression by soil microbiota (Adam et al., 2014). When defining the ISR, authors are very concerned not to mix this term with systemic acquired resistance, or priming (Martinez-Medina et al., 2016a). Pieterse et al. (2014) named it ISR when the induced resistance is triggered by a beneficial microbe and demonstrated to be SA-independent, and did not distinguish ISR and priming. Others (Van Wees et al., 2008; Martinez-Medina et al., 2016b) more specifically define defense priming being also systemic as ISR but can be induced by beneficial microbes, chemicals, pathogens, or other stress agents. In addition, priming always means that after its induction the enhanced defense response is independent from the presence of the inducer, i.e. it remains memorized *in planta*. This is not strictly required for ISR, so in this sense priming may be viewed as a special case of ISR.

There are two lines of plant defense. Pattern recognition receptors (PRRs) of plants first recognize microbial compounds, called microbe- or pathogen- or damage-associated molecular patterns (MAMPs, PAMPs or DAMPs) which induce a first line of defense, PAMP-triggered immunity or PTI (Boller and Felix, 2009). The second line of defense applies to successful pathogens which are able to combat effector-triggered immunity or ETI (Dodds and Rathjen, 2010). Over the years, several factors have been found to behave like MAMPs, including bacterial flagellin (Gómez-Gómez and Boller, 2000), lipopolysaccharides (LPS) (van Peer et al., 1991; Reitz et al., 2000), siderophores (Crowley, 2006; Beneduzi et al., 2012), antibiotics (Siddiqui and Shaukat, 2003b), biosurfactants (Ongena et al., 2007; Tran et al., 2007), or bacterial volatiles (Ryu et al., 2004; Naznin et al., 2014). Nevertheless, only few studies covered the involvement of MAMPs in nematode suppression in plants. When studying the suppression of the potato cyst nematode *Globodera pallida*, Reitz et al. (2000) showed in a split-root experiment that live cells or extracted LPS of *Rhizobium etli* strain G12 reduced reproduction of *G. pallida* by mediating ISR in potato plants.

Similarly, LPS of *P. fluorescens* strain WCS417 induced ISR in carnation and triggered an enhanced defense response after challenge inoculation with *Fusarium oxysporum* f. sp. *dianthi* (van Peer et al., 1991). In a recent study, Martinez-Medina et al. (2016) have tested local and systemic effects of the fungus *Trichoderma harzianum* T-78 on reproduction of *M. incognita*. They found a significant reduction of galls, shoot biomass and nematode fecundity both locally and systemically when plants were pre-inoculated with the fungus. More than that, this study showed the synergystic action of both salicylic acid (SA)- and jasmonic acid (JA)-mediated pathways. When plants were pre-treated with the fungus, a higher expression of SA-responsive marker genes, pathogenesis-related protein 1a (PR1) and Pathogenesis-related protein P6 (PR-P6) was observed in response to the RKN. Additionally, it was reported that JA-responsive marker genes Proteinase inhibitor II (PI II) and Multicystatin (MC) were down-regulated when nematodes were inoculated alone, but this down-regulation was suppressed when plants were pre-treated with the fungus. This effect was much stronger in systemic tissues suggesting that this JA-mediated pathway is not strongly expressed locally in the roots. Selim et al. (2014) have also found the involvement of both JA and SA in suppression of *M. javanica* by *T. harzianum T10*. The nematode fecundity was the lowest when the fungus was applied together with SA on the inducer side and it was preceded by an increase in the relative water content of the plant. The authors proposed that the plant resistance to nematodes appeared as a consequence of increased activity of peroxidase and phenoloxidase in plant tissues.

A split-root experiment was used to study the ability of bacterial antibiotics to mediate a plant response by ISR to RKN (Siddiqui and Shaukat, 2003b). More specifically, the effect of 2,4-diacetylphloroglucinol (DAPG) produced by *P. fluorescens* strain CHA0 against *M. javanica* was tested. It caused a reduction in the egg hatch and invasion of J2 in the roots, and increased J2 mortality. Similarly, Adam et al. (2014) have proved the ability of bacterial antagonists of fungal pathogens, *Bacillus subtilis* isolates Sb4-23, Mc5-Re2, and Mc2-Re2, to mediate ISR in tomato plants upon challenge inoculation with *M. incognita* and noted a significant reduction of plant infection by this nematode. The ISR was also responsible for a significant reduction of tomato root infection by *M. incognita* and *Pratylenchus penetrans*, upon challenging with the AMF *Funneliformis mosseae* (Vos et al., 2012). As an early MTI-response involving the jasmonate-linked 9-LOX-pathway has been unraveled in tomato plants inoculated with *F. mosseae* (Lopez-Raez et al., 2010)(Lopez-Raez et al. 2012), Schouteden et al. (2015) suggested that this response might be triggered during an early infection by root-knot nematodes as well. Recently, the class III chitinase gene VCH3 was found to be activated in *Glomus versiforme* colonized grape roots upon infection by *M. incognita* (Li et al., 2005). Hao et al. (2011) reported that *Glomus intraradices* colonization of grape (*Vitis* sp.) triggered activation of several genes after infection by the ectoparasitic nematode *Xiphinema index*.

Plants benefit from ISR by saving costs from activating the machinery of other defense-related genes that would otherwise affect plant growth and reproduction (Van Wees et al., 2008). Conversely, although most plants are in the state of ISR due to a vast number of rhizomicrobiota, sometimes the density of certain rhizomicrobes is not high enough to trigger ISR responses (Bakker et al., 2013). Nevertheless, if ignoring the latter and if considering the disadvantages of a direct antagonism of soil microbiota against PPN that are presented in previous sections, it can consequently be concluded that plant ISR responses represent an eminent and, yet, insufficiently explored weapon in the battle against phytonematodes.

Concluding remarks

It is irrefutable that PPN represent important members of plant devastators. Different strategies have been employed in the past decades to control yield losses that they cause, and yet, not many of them resulted in success. A microbial attachment to PPN represents a hot topic nowadays as an increasing number of examples has justified its contribution to a specific soil suppressiveness against PPN. Despite obligate parasitism is an important mode of action of many nematode-associated microbes to suppress PPN, we could see that there are many limitations for relying only on this mode of action. Many biocontrol agents that were found to successfully control phytonematodes *in vitro* or in the greenhouse could not show the same effect in the field. Some reasons for that are a weak competitive ability of such antagonists and failure to establish a high density that can significantly reduce PPN populations. On contrary, the intensification of studies on ISR-mediated suppression has shown its success in cases where a direct microbial antagonism failed. In the number of supporting examples, it has been understood that the core mechanism by which soilborne microbes counteract PPN relies on basal plant responses. However, many authors have not expanded their research beyond the split-root assays in this regard, assuming the involvement of the plant ISR without a molecular proof. Further studies are needed to unravel which genes and hormonal pathways are employed during plant basal defences against PPN that are triggered by nematode-associated microbiota. If the nematode population and species differences in microbial attachment are so pronounced, it would be intriguing to see how it affects and depends on plant ISR responses. A pursuit for specifically attached microbes that are more commonly associated with different nematode species and populations would aid their incorporation in biocontrol strategies against a broad spectrum of PPN to promote plant health.

References

Adam, M., Heuer, H., and Hallmann, J. (2014). Bacterial antagonists of fungal pathogens also control root-knot nematodes by induced systemic resistance of tomato plants. PLoS One *9*, e90402. doi: 10.1371/journal.pone.0090402.

Bais, H.P., L., W.T., Perry, L.G., S., G., and Vivanco, J.M. (2006). The role of root exudates in rhizosphere interactions with plants and other organisms. The Annual Review of Plant Biology *57*, 233-266. doi: 10.1146/annurev.arplant.57.032905.105159.

Baker, K.F., and Cook, R.J. (1974). Biological control of plant pathogens (San Francisco, USA: WH Freeman and Company). doi.

Bakker, P.A.H.M., Doornbos, R.F., Zamioudis, C., Berendsen, R.L., and Pieterse, C.M.J. (2013). Induced systemic resistance and the rhizosphere microbiome. Plant Pathology Journal *29*, 136-143. doi: 10.5423/PPJ.SI.07.2012.0111.

Ball, O.J.-P., Bernard, E.C., and Gwinn, K.D. (1997). Effect of selected *Neotyphodium lolii* isolates on root-knot nematode (*Meloidogyne marylandi*) numbers in perennial ryegrass. Paper presented at: Proceedings of the 50th New Zealand Plant Protection Conference. doi.

Banuelos, J., Alarcon, A., Larsen, J., Cruz-Sanchez, S., and Trejo, D. (2014). Interactions between arbuscular mycorrhizal fungi and *Meloidogyne incognita* in the ornamental plant *Impatiens balsamina*. Journal of Soil Science and Plant Nutrition *14*, 63-74. doi: 10.4067/S0718-95162014005000005.

Barron, G.L. (1977). The nematode-destroying fungi (Guelph, Ontario: Canadian Biological Publications Ltd.). doi.

Bell, N.L., Mtandavari, J.F., Maclean, P., Adam, K.H., Popay, A.J., Jones, R.J., Burch, G., Jones, R.D., and Fleetwood, D. (2016). Detection of invertebrate suppressive soils, and identification of a possible biological control agent for *Meloidogyne* nematodes using high resolution rhizosphere microbial community analysis. Frontiers in Plant Science *7*, 1-15. doi: 10.3389/fpls.2016.01946.

Beneduzi, A., Ambrosini, A., and Passaglia, L.M.P. (2012). Plant growth-promoting rhizobacteria (PGPR): Their potential as antagonists and biocontrol agents. Genetics and Molecular Biology *35*, 1044-1051.doi: 10.1590/S1415-475720 12000600020.

Bird, A.F., and Brisbane, P.G. (1988). The influence of *Pasteuria penetrans* in field soils on the reproduction of root-knot nematodes. Revue de Nématologie *11*, 75-81. doi.

Boller, T., and Felix, G. (2009). A renaissance of elicitors: Perception of microbe-associated molecular patterns and danger signals by pattern-recognition receptors. Annual Review of Plant Biology *60*, 379-406. doi: 10.1146/annurev.arplant.57.032905.105346.

Bordenstein, S.R., and Theis, K.R. (2015). Host biology in light of the microbiome: ten principles of holobionts and hologenomes. PLoS Biology *13*, e1002226. doi: 10.1371/journal.pbio.1002226.

Buchan, D., Moeskops, B., Ameloot, N., De Neve, S., and Sleutel, S. (2012). Selective sterilisation of undisturbed soil cores by gamma irradiation: Effects on free-living nematodes, microbial community and nitrogen dynamics. Soil Biology and Biochemistry *47*, 10-13. doi: 10.1016/j.soilbio.2011.12.014.

Bulgarelli, D., Rott, M., Schlaeppi, K., van Themaat, E.V.L., Ahmadinejad, N., Assenza, F., Rauf, P., Huettel, B., Reinhardt, R., Schmelzer, E., *et al.* (2012). Revealing structure and assembly cues for *Arabidopsis* root-inhabiting bacterial microbiota. Nature *488*, 91-95. doi: 10.1038/nature11336.

Castillo, J.D., Lawrence, K.S., Morgan-Jones, G., and Ramirez, C.A. (2010). Identification of fungi associated with *Rotylenchulus reniformis*. Journal of Nematology *42*, 313-318. doi.

Cetintas, R., and Dickson, D.W. (2004). Persistence and suppressiveness of *Pasteuria penetrans* to *Meloidogyne arenaria* Race 1. Journal of Nematology *36*, 540-549. doi.

Chen, Z.X., and Dickson, D.W. (1998). Review of *Pasteuria penetrans*: biology, ecology, and biological control potential. Journal of Nematology *30*, 313-340. doi.

Cook, R., and Lewis, G.C. (2001). Fungal endophytes and nematodes of agricultural and amenity grasses. In Biotic interactions in plant-pathogen associations, M.J. Jeger, and N.J. Spence, eds. (Wallingford, UK: CAB International), pp. 35-62. doi: 10.1079/9780851995120.0035.

Cooper, K.M., and Grandison, G.S. (1987). Effects of vesicular-arbuscular mycorrhizal fungi on infection of tamarillo by *Meloidogyne incognita* in fumigated soil. Plant Disease *71*, 1101-1106. doi.

Crowley, D.E. (2006). Microbial siderophores in the plant rhizosphere. In Iron nutrition in plants and rhizospheric microorganisms, L.L. Barton, and J. Abadia, eds. (Dordrecht, the Netherlands: Springer), pp. 169-198. doi: 10.1007/1-4020-4743-6_8.

Crump, D.H., and Kerry, B.R. (1987). Studies on the population dynamics and fungal parasitism of *Heterodera schachtii* in soil from a sugar-beet monoculture. Crop Protection *6*, 49-55. doi: 10.1016/0261-2194(87)90028-7.

Davies, K.G., Redden, M., and Pearson, T.K. (1994). Endospore heterogeneity in *Pasteuria penetrans* related to adhesion to plant-parasitic nematodes. Letters in Applied Microbiology *19*, 370-373. doi: 10.1111/j.1472-765X.1994.tb00478.x.

Davies, K.G., and Williamson, V.M. (2006). Host specificity exhibited by populations of endospores of *Pasteuria penetrans* to the juvenile andmale cuticles of *Meloidogyne hapla*. Nematology *8*, 475-476. doi.

Davies, K.G., Rowe, J.A., and Williamson, V.M. (2008). Inter- and intra-specific cuticle variation between amphimictic and parthenogenetic species of root-knot nematode (*Meloidogyne* spp.) as revealed by a bacterial parasite (*Pasteuria penetrans*). International journal for parasitology *38*, 851-859. doi: 10.1016/j.ijpara.2007.11.007.

Davies, K.G., and Curtis, R.H. (2011). Cuticle surface coat of plant-parasitic nematodes. Annual Review of Phytopathology *49*, 135-156. doi: 10.1146/annurev-phyto-121310-111406.

Davies, K.G., and Spiegel, Y. (2011). Biological control of plant-parasitic nematodes: Building coherence between microbial ecology and molecular mechanisms (Dordrecht, the Netherlands: Springer Science + Business Media). doi: 10.1007/978-1-4020-9648-8.

de la Pena, E., Echeverria, S.R., van der Putten, W.H., Freitas, H., and Moens, M. (2006). Mechanism of control of root-feeding nematodes by mycorrhizal fungi in the dune grass *Ammophila arenaria*. New Phytologist *169*, 829-840. doi: 10.1111/j.1469-8137.2005.01602.x.

de Medeiros, H.A., de Araújo Filho, J.V., de Freitas, L.G., Castillo, P., Rubio, M.B., Hermosa, R., and Monte, E. (2017). Tomato progeny inherit resistance to the nematode *Meloidogyne javanica* linked to plant growth induced by the biocontrol fungus *Trichoderma atroviride*. Scientific Reports *7*, 40216. doi: 10.1038/srep40216.

Dijksterhius, J., Veenhuis, M., Harder, W., and Nordbring-Hertz, B. (1994). Nematophagous fungi: physiological aspects and structure-function relationships. Advances in Microbial Physiology *36*, 111-139. doi: 10.1016/S0065-2911(08)60178-2.

Dodds, P.N., and Rathjen, J.P. (2010). Plant immunity: towards an integrated view of plant–pathogen interactions. Nature Reviews Genetics *11*, 539-548. doi: 10.1038/nrg2812.

Dong, L., Li, X., Huang, L., Gao, Y., Zhong, L., Zheng, Y., and Zuo, Y. (2014). Lauric acid in crown daisy root exudate potently regulates root-knot nematode chemotaxis and disrupts *Mi-flp-18* expression to block infection. Journal of Experimental Botany *65*, 131-141. doi: 10.1093/jxb/ert356.

Eilenberg, J., Hajek, A., and Lomer, C. (2001). Suggestions for unifying the terminology in biological control. BioControl *46*, 387-400. doi: 10.1023/A:1014193329979.

Elmi, A.A., West, C.P., Robbins, R.T., and Kirkpatrick, T.L. (2000). Endophyte effects on reproduction of root-knot nematode (*Meloidogyne marylandi*) and osmotic adjustment in tall fescue. Grass and Forage Science *55*, 166-172. doi.

Farnier, K., Bengtsson, M., Becher, P.G., Witzell, J., Witzgall, P., and Manduríc, S. (2012). Novel bioassay demonstrates attraction of the white potato cyst nematode *Globodera pallida* (Stone) to non-volatile and volatile host plant cues. Journal of Chemical Ecology *38*, 795-801. doi: 10.1007/s10886-012-0105-y.

Gómez-Gómez, L., and Boller, T. (2000). FLS2: An LRR receptor–like kinase involved in the perception of the bacterial elicitor flagellin in *Arabidopsis*. Molecular Cell *5*, 1003-1011. doi: 10.1016/S1097-2765(00)80265-8.

Hallmann, J., Quadt-Hallmann, A., Mahaffee, W.F., and Kloepper, J.W. (1997). Bacterial endophytes in agricultural crops. Canadian Journal of Microbiology *43*, 895-914. doi: 10.1139/m97-131.

Hallmann, J., Rodríguez-Kábana, R., and Kloepper, J.W. (1999). Chitin-mediated changes in bacterial communities of the soil, rhizosphere and within roots of cotton in relation to nematode control.

Soil Biology and Biochemistry *31*, 551-560. doi: 10.1016/S0038-0717(98)00146-1.

Hallmann, J., Quadt-Hallmann, A., Miller, W.G., Sikora, R.A., and Lindow, S.E. (2001). Endophytic colonization of plants by the biocontrol agent *Rhizobium etli* G12 in relation to *Meloidogyne incognita* infection. Phytopathology *91*, 415-422. doi: 10.1094/PHYTO.2001.91.4.415.

Handoo, Z.A. (1998). Plant-parasitic nematodes (USDA, ARS, Nematology Laboratory). doi.

Haney, C.H., Samuel, B.S., Bush, J., and Ausubel, F.M. (2015). Associations with rhizosphere bacteria can confer an adaptive advantage to plants. Nature Plants *1*, 15051. doi: 10.1038/nplants.2015.51.

Hao, Z., Fayolle, L., van Tuinen, D., Chatagnier, O., Li, X., Gianinazzi, S., and Gianinazzi-Pearson, V. (2011). Local and systemic mycorrhiza-induced protection against the ectoparasitic nematode *Xiphinema index* involves priming of defence gene responses in grapewine. Journal of Experimental Botany *63*, 3657-3672. doi: 10.1093/jxb/err313.

Hidalgo-Diaz, L., Bourne, J.M., Kerry, B.R., and Rodriguez, M.G. (2000). Nematophagous *Verticillium* spp. in soils infested with *Meloidogyne* spp. in Cuba: isolation and screening. International Journal of Pest Management *46*, 277-284. doi: 10.1080/09670870050206046.

Hol, W.H.G., and Cook, R. (2005). An overview of arbuscular mycorrhizal fungi–nematode interactions. Basic and Applied Ecology *6*, 489-503. doi: 10.1016/j.baae.2005.04.001.

Hussain, M., Zouhar, M., and Ryšánek, P. (2017). Effect of some nematophagous fungi on reproduction of a nematode pest, *Heterodera schachtii*, and growth of sugar beet. Pakistan Journal of Zoology *49*, 189-196. doi: 10.17582/journal.pjz/2017.49.1.189.196.

Inceoglu, O., van Overbeek, L.S., Falcao Salles, J., and van Elsas, J.D. (2012). Normal Operating Range of Bacterial Communities in Soil Used for Potato Cropping. Applied and Environmental Microbiology *79*, 1160-1170. doi: 10.1128/aem.02811-12.

Jaffee, B.A., Muldoon, A.E., and Tedford, E.C. (1992). Trap production by nematophagous fungi growing from parasitized nematodes. Phytopathology *82*, 615-620. doi.

Jia, C., Ruan, W.b., Zhu, M.-j., Ren, A.-z., and Gao, Y.-b. (2013). Potential antagnosim of cultivated and wild grass–endophyte associations towards *Meloidogyne incognita*. Biological Control *64*, 225-230. doi: 10.1016/j.biocontrol.2012.11.008.

Jorgenson, E.C. (1970). Antagonistic interaction of *Heterodera schachtii* Schmidt and *Fusarium oxysporum* (Woll.) on sugarbeets. Journal of Nematology 2, 393-398. doi.

Kerry, B. (1988). Fungal parasites of cyst nematodes. Agriculture, Ecosystems and Environment *24*, 293-305. doi.

Kiewnick, S., Neumann, S., Sikora, R.A., and Frey, J.E. (2011). Effect of *Meloidogyne incognita* inoculum density and application rate of *Paecilomyces lilacinus* strain 251 on biocontrol efficacy and colonization of egg masses analyzed by real-time quantitative PCR. Phytopathology *101*, 105-112. doi: 10.1094/phyto-03-10-0090.

Kloepper, J.W., and Beauchamp, C.J. (1992). A review of issues related to measuring colonization of plant roots by bacteria. Canadian Journal of Nematology *38*, 1219-1232. doi: 10.1139/m92-202.

Kloepper, J.W., Rodriguez-Kabana, R., McInroy, J.A., and Young, R.W. (1992). Rhizosphere bacteria antagonistic to soybean cyst (*Heterodera glycines*) and root-knot (*Meloidogyne incognita*) nematodes: identification by fatty acid analysis and frequency of biological control activity. Plant and Soil *139*, 75-84. doi: 10.1007/BF00012844.

Kluepfel, D.A., McInnis, T.M., and Zehr, E.I. (1993). Involvement of root-colonizing bacteria in peach orchard soils suppressive of the nematode *Criconemella xenoplax*. Phytopathology *83*, 1240-1245. doi.

Kumar, D., and Singh, K.P. (2006). Assessment of predacity and efficacy of *Arthrobotrys dactyloides* for biological control of root knot disease of tomato. Journal of Phytopatology *154*, 1-5. doi: 10.1111/j.1439-0434.2005.01047.x.

Le, H.T.T., Padgham, J.L., Hagemann, M.H., Sikora, R.A., and Schouten, A. (2016). Developmental and behavioural effects of the endophytic *Fusarium moniliforme* Fe14 towards *Meloidogyne graminicola* in rice. Annals of Applied Biology *169*, 134-143. doi: 10.1111/aab.12287.

Li, H.Y., Yang, G.D., Shu, H.R., Yang, Y.T., Ye, B.X., Nishida, I., and Zheng, C.C. (2005). Colonization by the arbuscular mycorrhizal fungus *Glomus versiforme* induces a defense response against the root-knot nematode *Meloidogyne incognita* in the grapevine (*Vitis amurensis* rupr.), which includes transcriptional activation of the class III chitinase gene *VCH3*. Plant and Cell Physiology *47*, 154-163. doi: 10.1093/pcp/pci231.

Liu, X., Xiang, M., and Che, Y. (2009). The living strategy of nematophagous fungi. Mycoscience *50*, 20-25. doi: 10.1007/s10267-008-0451-3.

Lopez-Llorca, L.V., Olivares-Bernabeu, C., Salinas, J., Jansson, H.-B., and Kolattukudy, P.E. (2002). Pre-penetration events in fungal parasitism of nematode eggs. Mycological Research *106*, 499-506. doi: 10.1017/s0953756202005798.

Lopez-Raez, J.A., Verhage, A., Fernandez, I., Garcia, J.M., Azcon-Aguilar, C., Flors, V., and Pozo, M.J. (2010). Hormonal and transcriptional profiles highlight common and differential host responses to arbuscular mycorrhizal fungi and the regulation of the oxylipin pathway. Journal of Experimental Botany *61*, 2589-2601. doi: 10.1093/jxb/erq089.

Manzanilla-Lopez, R.H., Esteves, I., Finetti-Sialer, M.M., Hirsch, P.R., Ward, E., Devonshire, J., and Hidalgo-Diaz, L. (2013). *Pochonia chlamydosporia*: Advances and challenges to improve its performance as a biological control agent of sedentary endo-parasitic nematodes. Journal of Nematology *45*, 1-7. doi.

Martinez-Medina, A., Flors, V., Heil, M., Mauch-Mani, B., Pieterse, C.M.J., Pozo, M.J., Ton, J., van Dam, N.M., and Conrath, U. (2016a). Recognizing plant defense priming. Trends in Plant Science *21*, 818-822. doi: 10.1016/j.tplants.2016.07.009.

Martinez-Medina, A., Pozo, M.J., Cammue, B.P.A., and Vos, C.M.F. (2016b). Belowground defence strategies in plants: the plant–*Trichoderma* dialogue. In Belowground Defence Strategies in Plants, C.M.F. Vos, and K. Kazan, eds. (Cham, Switzerland: Springer International Publishing), pp. 301-327. doi: 10.1007/978-3-319-42319-7_13.

Martínez-Medina, A., Fernández, I., Sánchez-Guzmán, M.J., Jung, S.C., Pascual, J.A., and Pozo, M.J. (2013). Deciphering the hormonal signalling network behind the systemic resistance induced by *Trichoderma harzianum* in tomato. Frontiers in Plant Science *4*. doi: 10.3389/fpls.2013.00206.

Martínez-Medina, A., Fernandez, I., Lok, G.B., Pozo, M.J., Pieterse, C.M., and Van Wees, S.C. (2016). Shifting from priming of salicylic acid- to jasmonic acid-regulated defences by *Trichoderma* protects tomato against the root knot nematode *Meloidogyne incognita*. New Phytologyst, 1-15. doi: 10.1111/nph.14251.

Martinuz, A., Schouten, A., and Sikora, R.A. (2013). Post-infection development of *Meloidogyne incognita* on tomato treated with the endophytes *Fusarium oxysporum* strain Fo162 and *Rhizobium etli* strain G12. BioControl *58*, 95-104. doi: 10.1007/s10526-012-9471-1.

Mazzola, M. (2004). Assessment and management of soil microbial community structure for disease suppression. Annual Review of Phytopathology *42*, 35-59. doi: 10.1146/annurev.phyto.42.040803.140408.

Meyer, S.L.F., Huettel, R.N., and Sayre, R.M. (1990). Isolation of fungi from *Heterodera glycines* and in vitro bioassays for their antagonism to eggs. Journal of Nematology *22*, 532-537. doi.

Meyer, S.L.F., and Wergin, W.P. (1998). Colonization of soybean cyst nematode females, cysts, and gelatinous matrices by the fungus *Verticillium lecanii*. Journal of Nematology *30*, 436-450. doi.

Moosavi, M.R., and Zare, R. (2012). Fungi as biological control agents of plant-parasitic nematodes. In Plant defence: Biological control, J.M. Mérillon, and K.G. Ramawat eds. (Heidelberg, Germany: Springer Science + Business Media), pp. 67-107. doi: 10.1007/978-94-007-1933-0_4.

Mukherjee, M., Horwitz, B.A., Berg, G., Mukherjee, P.K., Zachow, C., and Zeilinger, S. (2012). *Trichoderma*–plant–pathogen interactions: Advances in genetics of biological control. Indian Journal of Microbiology *52*, 522-529. doi: 10.1007/s12088-012-0308-5.

Naznin, H.A., Kiyohara, D., Kimura, M., Miyazawa, M., Shimizu, M., and Hyakumachi, M. (2014). Systemic resistance induced by volatile organic compounds emitted by plant growth-promoting fungi in *Arabidopsis thaliana*. PLOS ONE *9*, e86882. doi: 10.1371/journal.pone.0086882.

Nicol, J.M., Turner, S.J., Coyne, D.L., den Nijs, L., Hockland, S., and Maafi, Z.T. (2011). Current nematode threats to world agriculture. In Genomics and molecular genetics of plant-nematode interactions, J. Jones, G. Gheysen, and C. Fenoll, eds. (Dordrecht, the Netherlands: Springer), pp. 21-43. doi: 10.1007/978-94-007-0434-3_2.

Nigh, E.A., Thomason, I.J., and Van Gundy, S.D. (1980). Identification and distribution of fungal parasites of *Heterodera schachtii* eggs in California. Phytopathology *70*, 884-889. doi: 10.1094/Phyto-70-884.

Nyczepir, A.P., and Meyer, S.L.F. (2010). Host status of endophyte-infected and noninfected tall fescue grass to *Meloidogyne* spp. Journal of Nematology *42*, 151-158. doi.

Olatinwo, R., Becker, O.J., and Borneman, J. (2006). Suppression of *Heterodera schachtii* populations by *Dactylella oviparasitica* in four soils. Journal of Nematology *38*, 345-348. doi.

Ongena, M., Jourdan, E., Adam, A., Paquot, M., Brans, A., Joris, B., Arpigny, J.L., and Thonart, P. (2007). Surfactin and fengycin lipopeptides of

Bacillus subtilis as elicitors of induced systemic resistance in plants. Environmental Microbiology *9*, 1084-1090. doi: 10.1111/j.1462-2920.2006.01202.x.

Orion, D., Kritzman, G., Meyer, S.L.F., Erbe, E.F., and Chitwood, D.J. (2001). A role of the gelatinous matrix in the resistance of root-knot nematode (*Meloidogyne* spp.) eggs to microorganisms. Journal of Nematology *33*, 203-207. doi.

Ostendorp, M., and Sikora, R.A. (1989). Seed treatment with antagonistic rhizobacteria for the suppression of *Heterodera schachtii* early root infection of sugar beet. Revue de Nématologie *12*, 77-83. doi.

Pieterse, C.M.J., Zamioudis, C., Berendsen, R.L., Weller, D.M., Van Wees, S.C.M., and Bakker, P.A.H.M. (2014). Induced systemic resistance by beneficial microbes. Annual Review of Phytopathology *52*, 347-375. doi: 10.1146/annurev-phyto-082712-102340.

Reitz, M., Rudolph, K., Schröder, I., Hoffmann-Hergarten, S., Hallmann, J., and Sikora, R.A. (2000). Lipopolysaccharides of *Rhizobium etli* strain G12 act in potato roots as an inducing agent of systemic resistance to infection by the cyst nematode *Globodera pallida*. Applied and Environmental Microbiology *66*, 3515-3518. doi: 10.1128/AEM.66.8.3515-3518.2000.

Ryu, C.-M., Farag, M.A., Hu, C.-H., Reddy, M.S., Kloepper, J.W., and Paré, P.W. (2004). Bacterial volatiles induce systemic resistance in Arabidopsis. Plant Physiology *134*, 1017-1026. doi: 10.1104/pp.103.026583.

Saleh, H., and Sikora, R.A. (1984). Relationship between *Glomus fasciculatum* root colonization of cotton and its effect on *Meloidogyne incognita*. Nematologica *80*, 230-237. doi: 10.1163/187529284X00130.

Scher, F.M., Kloepper, J.W., Singleton, C.A., Zaleska, I., and Laliberte, M. (1988). Colonization of soybean roots by *Pseudomonas* and *Serratia* species: relationship to bacterial motility, chemotaxis, and generation time. Phytopathology *78*, 1055-1059. doi.

Schmidt, L.M., Hewlett, T.E., Green, A., Simmons, L.J., Kelley, K., Doroh, M., and Stetina, S.R. (2010). Molecular and morphological characterization and biological control capabilities of a *Pasteuria* ssp. parasitizing *Rotylenchulus reniformis*, the reniform nematode. Journal of Nematology *42*, 207-217. doi.

Schouteden, N., De Waele, D., Panis, B., and Vos, C.M. (2015). Arbuscular mycorrhizal fungi for the biocontrol of plant-parasitic nematodes: a review of the mechanisms involved. Frontiers in Microbiology *6*, 1280. doi: 10.3389/fmicb.2015.01280.

Schouten, A. (2016). Mechanisms involved in nematode control by endophytic fungi. Annual Review of Phytopathology *54*, 121-142. doi: 10.1146/annurev-phyto-080615-100114.

Selim, M.E., Mahdy, M.E., Sorial, M.E., Dababat, A.A., and Sikora, R.A. (2014). Biological and chemical dependent systemic resistance and their significance for the control of root-knot nematodes. Nematology *00*, 1-11. doi: 10.1163/15685411-00002818.

Sharon, E., Bar-Eyal, M., Chet, I., Herrera-Estrella, A., Kleifeld, O., and Spiegel, Y. (2001). Biological control of the root-knot nematode *Meloidogyne javanica* by *Trichoderma harzianum*. Phytopathol. *91*, 687-693. doi: 10.1094/PHYTO.2001.91.7.687.

Sharon, E., Chet, I., Viterbo, A., Bar-Eyal, M., Nagan, H., Samuels, G.J., and Spiegel, Y. (2007). Parasitism of *Trichoderma* on *Meloidogyne javanica* and role of the gelatinous matrix. European Journal of Plant Pathology *118*, 247-258. doi: 10.1007/s10658-007-9140-x.

Sharon, E., Chet, I., and Spiegel, Y. (2009). Improved attachment and parasitism of *Trichoderma* on *Meloidogyne javanica* in vitro. European Journal of Plant Pathology *123*, 291-299. doi: 10.1007/s10658-008-9366-2.

Siddiqui, I.A., and Shaukat, S.S. (2003a). Endophytic bacteria: prospects and opportunities for the biological control of plant-parasitic nematodes. Nematologia Mediterranea *31*, 111-120. doi.

Siddiqui, I.A., and Shaukat, S.S. (2003b). Suppression of root-knot disease by *Pseudomonas fluorescens* CHA0 in tomato: importance of bacterial secondary metabolite, 2,4-diacetylpholoroglucinol. Soil Biology and Biochemistry *35*, 1615-1623. doi: 10.1016/j.soilbio.2003.08.006.

Sikora, R.A., and Pocasangre, L. (2006). The concept of a suppressive banana plant: root health management with a biological approach. Paper presented at: Proceedings of the XVII ACROBAT International Congress (Joinville, Santa Catarina, Brazil). doi.

Sikora, R.A., Schäfer, K., and Dababat, A.A. (2007). Modes of action associated with microbially inducedin plantasuppression of plant-parasitic nematodes. Australasian Plant Pathology *36*, 124-134. doi: 10.1071/ap07008.

Sikora, R.A., Pocasangre, L.F., zum Felde, A., Niere, B., Vu, T.T., and Dababat, A.A. (2008). Mutualistic endophytic fungi and *in-planta* suppressiveness to plant parasitic nematodes. Biological Control *46*, 15-23. doi: 10.1016/j.biocontrol.2008.02.011.

Singh, K.P., Jaiswal, R.K., and Kumar, D. (2007). Nematophagous fungi associated with root galls of rice caused by *Meloidogyne graminicola* and its control by *Arthrobotrys dactyloides* and *Dactylaria brochopaga*. Journal of Phytopatology *155*, 193-197. doi: 10.1111/j.1439-0434.2007.01208.x.

Sitaramaiah, K., and Sikora, R.A. (1982). Effect of the mycorrhizal fungus *Glomus fasciculatus* on the host-parasite relationship of *Rotylenchulus reniformis* in tomato. Nematologica *28*, 412-419. doi: 10.1163/187529282X00213.

Spiegel, Y., Inbar, J., Kahane, I., and Sharon, E. (1995). Carbohydrate-recognition domains on the surface of phytophagous nematodes. Experimental Parasitology *80*, 220-227. doi: 10.1006/expr.1995.1027.

Stirling, G.R., and Mankau, R. (1978). Parasitism of *Meloidogyne* eggs by a new fungal parasite. Journal of Nematology *10*, 236-240. doi.

Stirling, G.R., Wilson, E.J., Stirling, A.M., Pankhurst, C.E., Moody, P.W., Bell, M.J., and Halpin, N. (2005). Amendments of sugarcane trash induce suppressiveness to plant-parasitic nematodes in a sugarcane soil. Australasian Plant Pathology *34*, 203. doi: 10.1071/ap05022.

Stirling, G.R. (2011). Suppressive biological factors influence populations of root lesion nematode (*Pratylenchus thornei*) on wheat in vertosols from the northern grain-growing region of Australia. Australasian Plant Pathology *40*, 416-429. doi: 10.1007/s13313-011-0059-5.

Stirling, G.R., Halpin, N.V., and Bell, M.J. (2011). A surface mulch of crop residues enhances suppressiveness to plant-parasitic nematodes in sugarcane soils. Nematropica *41*, 109-121. doi.

Stirling, G.R. (2014). Biological control of plant parasitic nematodes: Soil ecosystem management in sustainable agriculture, 2nd edn (Wallingford, UK: CAB International). doi: 10.1079/9781780644158.0000.

Stone, A.G., Scheuerell, S.J., and Darby, H.M. (2004). Suppression of soilborne diseases in field agricultural systems: Organic matter management, cover cropping, and other cultural practices. In Soil organic matter in sustainable agriculture, F. Magdoff, and R.R. Weil, eds. (Boca Raton: CRC Press), pp. 131-177. doi: 10.1201/9780203496374.ch5.

Tran, H., Ficke, A., Asiimwe, T., Höfte, M., and Raaijmakers, J.M. (2007). Role of the cyclic lipopeptide massetolide A in biological control of *Phytophthora infestans* and in colonization of tomato plants by *Pseudomonas fluorescens*. New Phytologist *175*, 731-742. doi: 10.1111/j.1469-8137.2007.02138.x.

van Loon, L.C., Bakker, P.A.H.M., and Pieterse, C.M.J. (1998). Induction and expression of PGPR-mediated induced resistance against pathogens. Biological control of fungal and bacterial plant pathogens - IOBC Bulletin *21*, 103-110. doi.

van Peer, R., Niemann, G.J., and Schippers, B. (1991). Induced resistance and phytoalexin accumulation in biological control of *Fusarium* wilt of carnation by *Pseudomonas* sp. strain WCS417r. Phytopathology *81*, 728-734. doi.

Van Wees, S.C.M., Van der Ent, S., and Pieterse, C.M.J. (2008). Plant immune responses triggered by beneficial microbes. Current Opinion in Plant Biology *11*, 443-448. doi: 10.1016/j.pbi.2008.05.005.

Vandenkoornhuyse, P., Quaiser, A., Duhamel, M., Le Van, A., and Dufresne, A. (2015). The importance of the microbiome of the plant holobiont. New Phytologist *206*, 1196-1206. doi: 10.1111/nph.13312.

Verdejo-Lucas, S., Ornat, C., Sorribas, F.J., and Stchiegel, A. (2002). Species of root-knot nematodes and fungal egg parasites recovered from vegetables in Almeria and Barcelona, Spain. Journal of Nematology *34*, 405-408. doi.

Veresoglou, S.D., and Rillig, M.C. (2011). Suppression of fungal and nematode plant pathogens through arbuscular mycorrhizal fungi. Biology Letters *8*, 214-217. doi: 10.1098/rsbl.2011.0874.

Vos, C.M., Tesfahun, A.N., Panis, B., De Waele, D., and Elsen, A. (2012). Arbuscular mycorrhizal fungi induce systemic resistance in tomato against the sedentary nematode *Meloidogyne incognita* and the migratory nematode *Pratylenchus penetrans*. Applied Soil Ecology *61*, 1-6. doi: 10.1016/j.apsoil.2012.04.007.

Westphal, A., and Becker, J.O. (1999). Biological suppression and natural population decline of *Heterodera schachtii* in a California field. Phytopathology *89*. doi: 10.1094/PHYTO.1999.89.5.434.

Westphal, A. (2005). Detection and description of soils with specific nematode suppressiveness. Journal of Nematology *37*, 121-130. doi.

Westphal, A., Pyrowolakis, A., Sikora, R.A., and Becker, J.O. (2011). Soil suppressiveness against Heterodera schachtii in California cropping areas. Nematropica *41*, 161-171. doi.

CHAPTER 6

Apple Replant Disease: Causes and Mitigation Strategies

Traud Winkelmann[1]*, Kornelia Smalla[2], Wulf Amelung[3], Gerhard Baab[4], Gisela Grunewaldt-Stöcker[5], Xorla Kanfra[2], Rainer Meyhöfer[5], Stefanie Reim[6], Michaela Schmitz[7], Doris Vetterlein[8,9], Andreas Wrede[10], Sebastian Zühlke[11], Jürgen Grunewaldt[12], Stefan Weiß[1] and Michael Schloter[13, 14]

[1]Institute of Horticultural Production Systems, Section Woody Plant and Propagation Physiology, Leibniz Universität Hannover, Herrenhäuser Str. 2, D-30419 Hannover, Germany
[2]Institute for Epidemiology and Pathogen Diagnostics, Julius Kühn-Institut, Messeweg 11/12, D-38104 Braunschweig, Germany
[3]Institute of Crop Science and Resource Conservation, Division Soil Science, Universität Bonn, Nussallee 13, D-53115 Bonn, Germany
[4]Competence Center of Horticulture, DLR Rheinpfalz, Campus Klein Altendorf 2, D-53359 Rheinbach
[5]Institute of Horticultural Production Systems, Section Phytomedicine, Leibniz Universität Hannover, Herrenhäuser Str. 2, D-30419 Hannover, Germany
[6]Institute for Breeding Research on Fruit Crops, Julius Kühn-Institut, Pillnitzer Platz 3a, D-01326 Dresden, Germany
[7]Department of Applied Science, Hochschule Bonn-Rhein-Sieg, Von-Liebig-Str. 20, D-53359 Rheinbach, Germany
[8]Institute of Agricultural and Nutritional Sciences, Soil Sciences, Martin-Luther Universität Halle-Wittenberg, Von-Seckendorff-Platz 3, D-06120 Halle/Saale, Germany
[9]Department of Soil System Science, Helmholtz Center for Environmental Research, UFZ, Theodor-Lieser-Straße 4, D-06120 Halle/Saale, Germany
[10]Department of Horticulture, Landwirtschafts-kammer Schleswig-Holstein, Thiensen 16, D-25373 Ellerhoop, Germany
[11]Institute of Environmental Research (INFU), Department of Chemistry and Chemical Biology, Chair of Environmental Chemistry and Analytical Chemistry, Technische Universität Dortmund, Otto-Hahn-Str. 6, D-44221 Dortmund, Germany
[12]Institute of Plant Genetics, Unit Molecular Plant Breeding, Leibniz Universität Hannover, Herrenhäuser Str. 2, D-30419 Hannover, Germany
[13]Research Unit for Comparative Microbiome Analysis, Helmholtz Zentrum München, Ingolstädter Landstr. 1, D-85764 Neuherberg, Germany
[14]Chair for Soil Science, Research Department Ecology and Ecosystem Management, Technische Universität München, Emil-Ramann-Straße 2, D-85354 Freising, Germany

*traud.winkelmann@zier.uni-hannover.de

DOI: https://doi.org/10.21775/9781912530007.06

Abstract

After replanting apple (*Malus domestica* Borkh.) on the same site severe growth suppressions, and a decline in yield and fruit quality are observed in all apple producing areas worldwide. The causes of this complex phenomenon, called apple replant disease (ARD), are only poorly understood up to now which is in part due to inconsistencies in terms and methodologies. Therefore we suggest the following definition for ARD: ARD describes a harmfully disturbed physiological and morphological reaction of apple plants to soils that faced alterations in their (micro-) biome due to the previous apple cultures. The underlying interactions likely have multiple causes that extend beyond common analytical tools in microbial ecology. They are influenced by soil properties, faunal vectors, and trophic cascades, with genotype-specific effects on plant secondary metabolism, particularly phytoalexin biosynthesis. Yet, emerging tools allow to unravel the soil and rhizosphere (micro-) biome, to characterize alterations of habitat quality, and to decipher the plant reactions. Thereby, deep insights into the reactions taking place at the root rhizosphere interface will be gained. Counteractions are suggested, taking into account that culture management should emphasize on improving soil microbial and faunal diversity as well as habitat quality rather than focus on soil disinfection.

Introduction

Apple replant disease (ARD) affects plant propagation in nurseries as well as apple production worldwide by strongly reducing plant growth as well as fruit yield and quality (Mazzola and Manici, 2012; Manici et al., 2013). The disease occurs after repeated replanting of apple at the same site. On ARD soils, over the lifetime of an apple orchard, a 50 % reduced profitability has been estimated due to later and less fruit bearing of the affected trees (Mazzola, 1998; Van Schoor et al., 2009). Problems by ARD are increasing recently, mainly due to the concentration of tree nurseries to certain regions, such as the Pinneberg region in Germany or Pistoia in Italy, as well as due to the concentration of apple orchards in the respective fruit production areas. In addition, to achieve higher planting densities, the use of dwarf rootstocks results in a shorter life span of these orchards and in more frequent replanting (St. Laurent et al., 2010; Volk et al., 2015). A rapid improvement of this situation is unlikely, as installation costs increase (frost protection, irrigation) and areas for crop rotation become increasingly scarce due to alternative usage for industry, energy plants, or other purposes.

Various definitions of the term "replant disease" or related phrases such as "replant problem", "soil sickness" or "soil fatigue" exist (e.g. Klaus, 1939; Hoestra, 1968; Utkhede, 2006). According to Utkhede (2006) "replant problems" include both, abiotic and biotic factors which suppress plant growth, whereas "replant disease" comprises only all biotic factors. "Soil sickness" is used in cases, where the causes of the reduced growth are unknown or uncertain (Savory, 1966), thus excluding nematode damage (Spethmann and Otto, 2003). Here, we suggest the following definition of ARD: *ARD describes a harmfully disturbed physiological and morphological reaction of apple plants to soils that faced alterations in their (micro-) biome due to previous apple cultures.*

ARD is characterized by the specificity to the species *Malus domestica* (although cross reactions with other Rosaceae have been observed) and a persistence for decades (Savory, 1966). ARD is reversible after transplanting into virgin or healthy soil. Replant disease has also been reported for other plants, and especially members of the Rosaceae, such as cherry, peach, strawberry, rowan, and rose, are prone to it. In this review, we summarize current knowledge on the causes of ARD and critically evaluate current lines of research to develop mitigation strategies.

Etiology and causes of ARD

The species specificity implies that ARD has its origin in the apple plant-soil interface: Based on root exudates (Börner, 1959; Wittenmayer and Szabó, 2000; Hofmann et al., 2009) or/and by decomposition products of dead apple plant material changes in the biomes of the rhizosphere and of the soil are induced. A recent study addressed the composition of root deposits, thus analyzing the important root soil interface, of different apple rootstocks for the first time (Leisso et al., 2017). The rhizodeposits were found to be highly dynamic and influenced by growth conditions, rootstock genotype and bacterial communities of the rhizosphere. Also intraspecific allelopathy cannot be excluded in this context, but the persistence of ARD over decades would suggest that the involved toxic or deleterious substances are extraordinarily stable and bioactive. It is well accepted that soils may store or cycle certain molecules such as carbohydrates, lipids or proteins for years to decades (e.g. Wiesenberg et al., 2004; Derrien et al., 2006; Amelung et al., 2008). However, this has not been proven for root exudates, as usually these molecules are quickly transformed by rhizomicrobial respiration before being adsorbed or bound to soil minerals, which would make them inert against microbial degradation (e.g. Oades, 1988; Kuzyakov and Larionova, 2005; von Lützow et al., 2006; Kögel-Knabner and Amelung, 2014). If they do get bound, then it is questionable whether such compounds really maintain their bioavailability. From experiments with pollutants like polycyclic aromatic hydrocarbons, pesticides, or antibiotics, we know that with increasing contact time, newly added compounds become sequestered, thus rapidly losing their desorbability, bioavailability, and potential effectiveness on soil microbes (e.g. Hatzinger and Alexander, 1995; Lueking et al., 2000; Ciglasch et al., 2008; Rosendahl et al., 2012). It therefore appears truly unlikely that toxic compounds can explain ARD persistence for years. However, they may still trigger initial shifts in microbiome composition.

Shifts in bacterial and fungal communities in soil as a trigger for ARD development

Previous studies allow the conclusion that ARD is assumed to be a disease-complex (Figure 1) which is influenced by the soil type and the climate of the respective site. The fact that soil disinfection leads to the restored regular plant growth clearly points to biotic causal factors (e.g. Mai and Abawi, 1981; Mazzola, 1998; Yim et al., 2013; Spath et al., 2015). In earlier studies on ARD soil microorganisms, various genera differing in dominance between sites were suggested to be involved in the disease complex. These are the oomycetes *Pythium* and *Phytophthora*, the fungi *Cylindrocarpon* and *Rhizoctonia* as well as actinomycetes and other bacterial genera like *Bacillus* and *Pseudomonas* (Jaffee et al., 1982; Otto et al., 1994; Mazzola, 1998; Utkhede, 2006; Tewoldemedhin et al., 2011). Also evidence for an involvement of root endophytic fungi (e.g. *Cylindrocarpon*-like fungi) in apple plant growth reduction on ARD affected soils was presented (Manici et al., 2013, 2018). More recently, numerous studies on microbial community analyses of ARD soils were published, strongly fostered by new sequencing technologies (Rumberger et al., 2007; Tewoldemedhin et al., 2011; Yim et al., 2013, 2015, 2016; Sun et al., 2014; Caputo et al., 2015; Franke-Whittle et al., 2015; Mazzola et al., 2015; Hewavitharana and Mazzola, 2016; Nicola et al., 2017). These total DNA based analyses confirmed previous data on changes in the microbial community composition in replant soils. Obviously, the soil-inherent microbial diversity, the plant species, and the plant growth stage influence the microbiome in the rhizosphere of apple plants. Differences are observed in the bacterial and fungal community composition in ARD affected and in healthy soils from the same site (e.g. Franke-Whittle et al., 2015). However, up to now it is unclear, whether the missing or additional microorganisms as well as shifts in abundances caused ARD or

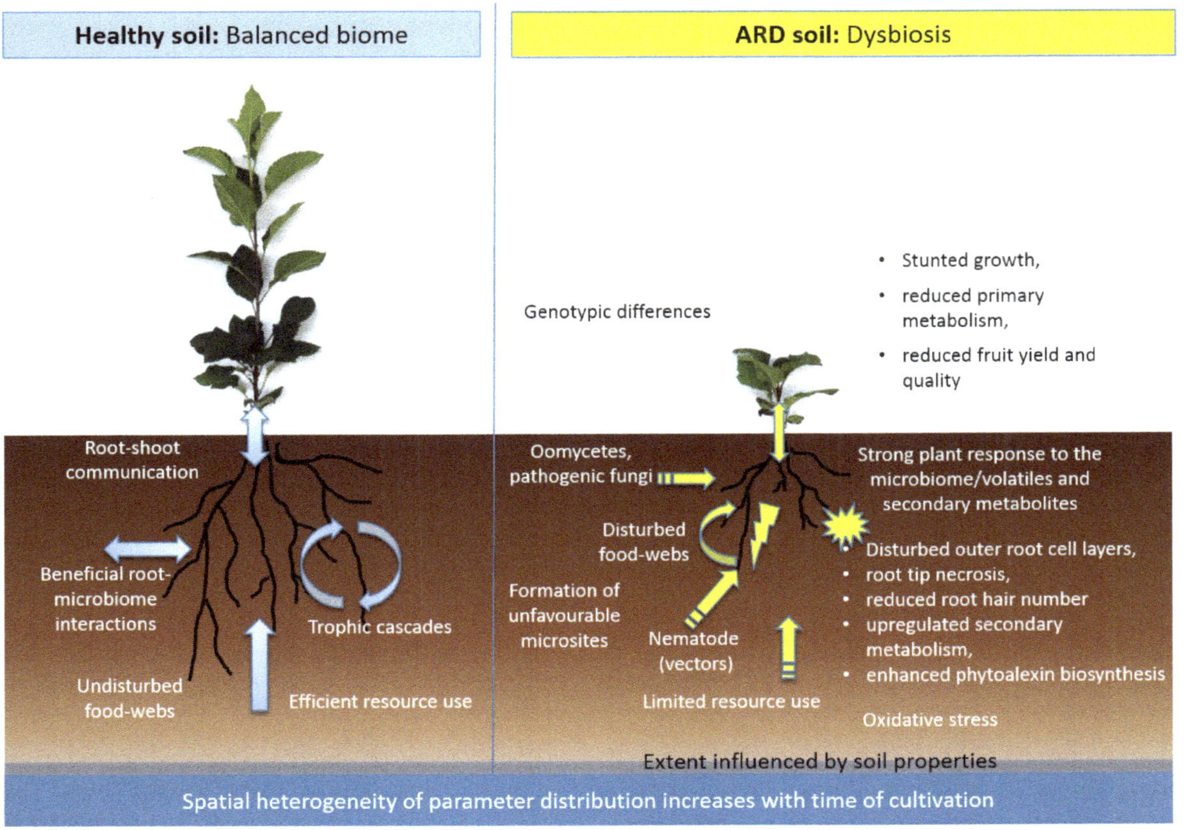

Figure 1. Apple replant disease (ARD) has multiple causes with a strong impact of dysbiosis regarding the microbiome and is influenced by soil properties, faunal vectors, and trophic cascades, with genotype-specific reactions.

occurred as a result of ARD. The analysis of amplicon sequencing data of microorganisms in ARD soil after treatment with heat, gamma irradiation, Basamid® or biofumigation revealed numerous bacterial and fungal populations with significantly increased abundance (responders) compared to that in untreated ARD soil. Although the 16S rRNA gene and ITS sequence data need a careful interpretation, notably, many of the responders belonged to taxa of which strains with plant beneficial traits or antagonistic activity were described, such as *Burkholderia* ssp., *Arthrobacter* ssp., *Podospora* ssp. or *Penicillium* ssp. (Franke-Whittle et al., 2015; Mazzola et al., 2015; Yim et al., 2016, 2017). However, the analysis of diversity revealed on the base of 16S rRNA gene fragments or ITS fragments amplified from DNA directly extracted from soil (total community DNA - TC-DNA) is limited in its resolution down to the genus or species level depending on the used primer systems. But, for many species which colonize the rhizosphere, it is well known that traits differ on a strain specific level and bacteria of the same species can act as phytopathogens or plant growth promoting bacteria. Even the same strain can change the gene expression pattern depending on its environment and thus develops different interaction patterns with plant roots. Thus, analyses of the functions and the expression pattern of genes of interest provided by the below-ground microbiome are needed to improve our mechanistic understanding on the role of microbes in ARD development.

In upcoming studies oomycetes need to be included in the analysis by molecular tools as the primers targeting fungal ITS do not amplify oomycetes and thus these data are still missing. Furthermore, qPCR systems need to be established to determine the changes in abundance of potential harmful or beneficial microorganisms in the apple rhizosphere microbiome in response to different soil management treatments.

Soil fauna affecting ARD

Besides nematodes (see below), important soil mesofauna groups are Collembolans and soil mites, which play an important role in soil food webs as decomposers, plant parasites, microbivores as well as predators (Hopkin, 1997; Wardle, 2006, 2013; Walter and Proctor, 2013). Both groups of organisms are neither discussed in the context of causing agents nor considered at all in the current ARD literature (Utkhede, 2006; Mazzola and Manici, 2012; Vukicevich et al., 2016). Nevertheless, there is much evidence in the literature that soil mesofauna affects not only bacterial and fungal communities in the rhizosphere, i.e. by selective feeding on pathogenic or non-pathogenic microorganisms (Lartey et al., 1994; Sabatini and Innocenti, 2000, 2001; Innocenti et al., 2009; Böllmann et al., 2010), but also promotes mycorrhizal fungi (Steinaker and Wilson, 2008; Kanters et al., 2015) and other beneficial microorganisms (Lartey et al., 1994). Having in mind that Collembolans and soil mites are also strongly affected by land management (fertilizer, water, soil structure) (Schrader et al., 1997; Larsen et al., 2004; Innocenti et al., 2011; Roy et al., 2014) and pesticide use (Frampton, 2002; San Miguel et al., 2008; Chelinho et al., 2014), it would not be surprising that cascading trophic effects driven by consumers either directly or indirectly influence the etiology of ARD. So far, own results indicate decreased abundance of Collembola and soil mites, as well as shifts in Collembolan species composition in ARD compared to healthy soils (Meyhöfer et al., unpublished data).

Many studies suggested a role of root lesion nematodes in the ARD development (Hoestra and Oostenbrink, 1962; Dunn and Mai, 1972; Mai and Abawi, 1981; Jaffee et al., 1982) citing notably uneven distribution pattern of *Pratylenchus penetrans* in apple orchards (Mai and Abawi, 1978; Jaffee et al., 1982; Mai et al., 1994). Yim et al. (2013) inactivated nematodes in ARD affected soils by heat treatment and could prove that apple plants grew significantly better in heat-treated ARD soil compared to the untreated ARD soil, confirming the role of nematodes in ARD development. However, nematicide applications in affected orchards were inefficient to enhance apple growth (Hoestra and Oostenbrink, 1962; Covey Jr et al., 1979; Caruso et al., 1989; Mazzola, 1998). Furthermore, the low frequency of endoparasitic nematodes in roots did not give evidence for a contribution to growth reduction in apple in ARD affected soils (Manici et al., 2013). Nevertheless, root lesions induced by nematodes can cause synergistic damage to apple by acting in combination with some notable pathogenic fungi or oomycetes such as *Rhizoctonia, Phytophthora, Cylindrocarpon,* and *Pythium* (Mazzola, 1998). Furthermore, a high abundance of nematodes feeding on microbes can modify the microbial community by altering the relative abundance of populations (Djigal et al., 2004; Hai-Feng et al., 2014; Gebremikael et al., 2016) thus causing a significant reduction of microbes that induce plant growth promotion. Indirectly, nematodes might contribute to ARD by dissemination of microbes (Freckman, 1988; Wang and McSorley, 2005) or activation of specific microbial growth by the release of growth limiting nutrients (Wang and McSorley, 2005). Recent findings by Adam et al. (2014) and Elhady et al. (2017) confirm specific bacteria and fungi to be attached to infective stages of *Meloidogyne incognita* and *P. penetrans* in different soil types indicating an ecological role of the association. Four way interaction between fungi, oomycetes, bacteria and nematodes was supposed to increase the ARD severity when these organisms were present at the same time (Utkhede et al., 1992; Mazzola and Mullinix, 2005).

Soil properties affecting ARD

Soil properties might modulate the degree of ARD observed in plants, if not even causally affecting it (von Bronsart, 1949). In general, sandy/light soils have often been observed to be more prone to ARD than loamy soils (http://www.leicesters.co.nz/specific-apple-replant-disease/). Own observations additionally confirm that areas with high groundwater levels or extended periods of water logging are less conducive for ARD most likely due to interfering with the development of pathogenic aerobic communities.

Von Bronsart (1949) stated that physical soil conditions like compaction, loss of specific pore sizes, dryness in macropores or stagnant water conditions above a compacted plough pan, may hardly be seen as major cause of ARD, because the latter is plant-specific while most soil-related effects are not. Nevertheless, physical soil conditions may drive the survival and competitiveness of phytopathogenic or beneficial nematodes and also of microbial communities, i.e., they might affect the intensity and duration of ARD symptoms. And indeed, ARD has often been observed to occur heterogeneously at a given site, as do many soil properties (Bogena et al., 2010; Gebbers and Adamchuk, 2010; Herbst et al., 2012). Mazzola and Manici (2012) concluded that abiotic factors may exacerbate ARD but do not appear to function as the primary cause of the disease.

Among chemical soil properties influencing ARD severity, pH is certainly a master variable affecting nutrient availability, microbial diversity and microbial nutrient mobilization and immobilization. Some authors reported that in soils with low pH values (around 4-4.5) ARD problems are less pronounced (Jonkers et al., 1980; Utkhede, 2006), while others reported that low pH values make soils prone to replant disease problems (Willet et al., 1994). Additionally, the role of pH on ARD severity is obviously genotype dependent (Fazio et al., 2012).

With variations in pH, nutrient availability changes, and there has been focus on micronutrient controls on ARD expression, mainly related to Zn, Fe, and Mo (von Bronsart, 1949; Fan et al., 2010; Fazio et al., 2012). To avoid additional nutrient limitations, apple replanting should consider respective fertilizer recommendations. These may include, for instance, the use of selected micronutrients such as zinc while possibly excluding or at least very carefully operating with others like boron (Tukey et al., 1984). Adding mineral fertilizer, e.g., with P, (Sewell et al., 1988), as well as managing ground cover by adding compost or even biochar, may induce increased biocontrol properties of soils including a reduction of nematodes causing root lesions. Fertilization may thus affect ARD severity in nutrient-limited soils (van Schoor et al., 2009), but frequently failed to replace fumigation for replanting in temperate climates (St. Laurent et al., 2008; Mazzola and Manici, 2012; Glisczynski et al., 2016; Peruzzi et al., 2017).

To future study the role of soil in modulating the degree of ARD in plants, it will be inevitable to map spatial ARD heterogeneity in different orchards and to correlate it with spatial patterns of soil properties that likely also include subsoil properties. The finding of causal interactions between soil properties and ARD, however, is often hampered by insecurity related to sampling rhizosphere soil. Rhizosphere soil is commonly obtained by a vigorous shaking of the uprooted root system. This enables neither to differentiate between affected and non-affected root areas, nor to define precisely the distance from the root surface. As in general less material is obtained from affected roots compared to healthy root material, a dilution of effects might occur using total root sampling approaches.

Plant reactions to ARD

The level of susceptibility differs significantly between individual apple genotypes. Fully resistant genotypes were never observed yet, but less susceptible and/or tolerant genotypes can be found for different species of the genus *Malus* (Isutsa and Merwin, 2000). Symptoms of ARD are expressed early after the first contact with ARD affected soil and include belowground a root browning and blackening, root tip necrosis, reduced number of root hairs and destroyed outer root cell layers

(Caruso et al., 1989; Yim et al., 2013, Grunewaldt-Stöcker, unpublished data, Figure 2).

Aboveground plant parts show stunted or rosette growth (Caruso et al., 1989; Mazzola, 1998; Mazzola and Manici, 2012; Yim et al., 2013; Atucha et al., 2014; Emmett et al., 2014). These severe disorders result in a dramatically reduced plant biomass, fruit yield, as well as fruit size and flavor (Mazzola and Manici, 2012; Liu et al., 2014). The molecular and physiological reactions of apple plants to ARD soils resulting in these morphologically visible symptoms were only recently subject of first in-depth studies. The accumulation of phenolic compounds as antioxidants in roots and shoots under ARD, points to oxidative stress (reactive oxygen species) (Henfrey et al., 2015), and may be a consequence of plant damage from ARD induced plant secondary metabolites. Changes in patterns of phenolic compounds, like phloridzin and phloretin, benzoic acid and rutin (Börner, 1959; Hofmann et al., 2009; Yin et al., 2016, 2017; Leisso et al., 2017) could be the result or reason of ARD. The exudation of these compounds may affect the complete soil microbiome or parts of it.

In addition, the abundance of antioxidative enzymes, such as peroxidases increased significantly four weeks after planting young apple rootstocks in ARD affected soil, whereas synthesis of these enzymes was lower in plants grown in gamma irradiated soil (Schmitz et al., unpublished results). Peroxidases oxidize hydrogen donors at the expense of peroxides. They are highly specific for hydrogen peroxide, but they accept a wide range of other hydrogen donors, including polyphenols. The higher activity of peroxidase could promote the oxidation of phenols into the antioxidative polyphenols in the roots (Ayyagari et al., 1996), which may lead to the visible browning and blackening symptoms as well as root tip necrosis.

After infection of roots of apple seedlings by *Pythium ultimum*, one of the potential causal agents of ARD, an upregulation of the expression of genes involved in the secondary metabolism occurred as well as differential expression of genes in plant

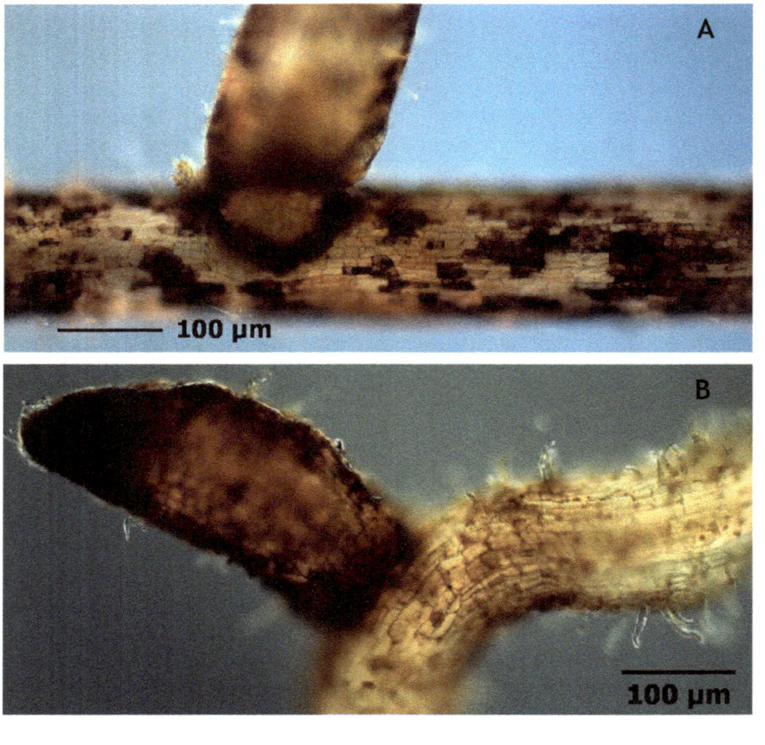

Figure 2. Cell necroses and blackening in the outer tissue layers of a branching fine root (A), and root tip necrosis (B) of apple rootstock *Malus domestica* M26, grown for two weeks in ARD affected soil.

hormone metabolism (Shin et al., 2014, 2016; Zhu et al., 2014). Comparative transcriptomic studies of roots of the sensitive rootstock M26 grown either in ARD or gamma irradiated ARD soils revealed several differences in the expression of genes involved in stress responses (Weiß et al., 2017a, b). Further, when grown in ARD soil the plants reacted with an upregulation of expression of genes of the secondary metabolism, especially concerning the phytoalexin biosynthesis. Also, the corresponding phytoalexin products, i.e. biphenyls and dibenzofurans were detected in relatively high concentrations (Weiß et al., 2017b). A more detailed understanding of the molecular interplay of apple plants and their microbiome in healthy and ARD affected soils is urgently needed to define causes and consequences of ARD for plants and microbes (Manici et al., 2017).

The role of spatial distribution of relevant parameters

There are several observations indicating that the ARD causing agent, whatever it is, lacks mobility. Hoestra (1968) already reported in 1968, that ARD affects the apple tree in the first years of planting, thereafter the roots grow into deeper soil layers less impaired by ARD. He showed in growth experiments with soil extracted from different depths that ARD was mainly observed in 0-15 and 15-30 cm soil depth. As also many soil properties and functions are heterogeneous under field conditions (Bogena et al., 2010; Gebbers and Adamchuk, 2010; Herbst et al., 2012), the patchiness of ARD related growth depression in the field as reported above, likewise hints in the same direction. Interestingly, ARD is induced more rapidly if the site is replanted frequently (nurseries) compared to sites permanently used for apple production. Frequent replanting is associated with more frequent mixing of soil due to uprooting and soil cultivation.

Restricted mobility of ARD causing agents, at least within the root system was also confirmed in a recent split-root experiment (Figure 3) of Lucas et al. (unpublished data). Apple plants grown in split-root systems with different combinations of ARD soil, sterilized ARD soil or control soil (same site but never planted with apple) gained no reduction in shoot growth if half of the root system had access to soil not affected by ARD. The spatial separation is obviously crucial as simple dilution of the ARD soil by sterilized or control soil did not lead to comparable results (Hoestra, 1968; Jaffee et al., 1982; van Schoor et al., 2009; Tewoldemedhin et al., 2011; Spath et al., 2015). The split-root experiment of Lucas et al. (submitted) clearly showed that ARD is not systemic. Bacterial and fungal community composition in the rhizoplane and rhizosphere of the same plants differed significantly between the compartments containing ARD soil and those containing sterilized or control soil. However, some populations were only detected in the sterilized soil if the neighboring compartment contained ARD soil.

Figure 3. Root growth of *Malus domestica* M26 in a splitroot experiment. (Photo taken by Maik Lucas).

Further observations from our group (Zickenrott et al., unpublished data) indicate that apple plant roots avoid ARD soil patches, if given a choice. The mechanisms behind this are currently not known.

Assessment of mitigation strategies for ARD

Crop rotation is the first and oldest way to mitigate or circumvent ARD (Mazzola and Gu, 2001), but this is strongly limited or even not possible due to high investments in orchard infrastructure, for-instance in fences, hail nets, wells, pipes and more technology for irrigation. However, the main obstacle is the lack of areas for rotation in the production centers. Soil fumigation by chemicals is no longer possible in many countries due to the phase-out of the ecologically harmful fumigants. Biofumigation (Brown et al., 1991) using the incorporation of Brassicaceae plants or seed meal has been suggested as a counteraction and has shown first promising results (Mazzola et al., 2001, 2007, 2015; Yim et al., 2016, 2017), but cannot fully restore plant growth in most cases. The authors could prove that the application of *Brassica napus* seed meal amendments resulted in an increased abundance of *Actinomycetes,* e.g. *Streptomycetes,* and *Pseudomonas* in soil, bacterial groups being known for their high contribution to biocontrol of phytopathogens (Mazzola et al., 2007). The observed effects of applied *B. napus* seed meal were, however, variable and depended on the time of application, the concentration applied and the content of glucosinolates of the meal.

Further, several studies in the past have investigated the impact of fungicides like difenconazole or metalaxyl on the growth of apple trees in soils with replant disease symptoms (Mazzola, 1998). Although positive effects were obvious the issue of sustainability is questionable as a continuous application is needed. Because of the small specificity of the compounds other non-target populations like beneficial fungi might be affected with non-intended side effects.

Steam disinfection of soils is theoretically possible but too energy and time consuming and still fraught with technical problems, as demonstrated in current experiments in German nurseries. The costs of disinfecting soils with steam are 3-4 times higher than using chemicals (Nitt et al., 2015). Interestingly, the intercropping with *Tagetes*, conventionally used against nematodes, revealed increased growth of apple in two ARD soils, both in a bio-test as well as in field trials (Yim et al., 2017).

The idea to change more than the abundance of one microbial strain becomes more and more popular as it is well accepted that different microbial traits might contribute to overcome replant disease in soil, and that soil microbial diversity is strongly altered by replant disease (Sun et al., 2014; Berg et al., 2017). Already 25 years ago, Utkhede and Smith (1992) reported the promotion of apple tree growth and fruit production in a former ARD soil after inoculation with a strain of *Bacillus subtillis,* which showed biocontrol activities against various pathogens. The authors could prove that the inoculation procedure was more effective than a classical formalin fumigation, mainly, as it did not only increase shoot growth and the cross-sectional trunk area but resulted in higher yields, too. This concept of disease suppression via the inoculation of biocontrol microbes was further followed up, mainly as several authors could prove a high abundance of *Rhizoctonia spp.* from ARD soils (Mazzola, 1997). In 2007, the same author published data on the manipulation of rhizosphere bacterial communities to induce suppressive soils (Mazzola, 2007). Crop rotation, including wheat cultivation after apple growth, reduced the susceptibility of soils for ARD, and correspondingly an increase of fluorescent pseudomonads in the soil was observed (Mazzola et al., 2002). Therefore, it was suggested to use selected *Pseudomonas* strains of the species *P. fluorescens* or *P. putida* with biocontrol properties against *Rhizoctonia* for inoculation (Mazzola et al., 2002). These approaches seem to be promising, since the use of chemical substances can be avoided. However, it needs to be taken into account that microbe-based inoculation strategies need to consider on the one hand the potential risk of the inoculum for the environment. For example, *P. putida* has been recently classified into risk class II according to the German biosafety level, as several severe cases of infections of humans *with P. putida* have been reported (Carpenter et al., 2008). On the other hand, inoculation-based approaches often do not result in the expected outcome as the inoculated microbes did not establish well in soil and were outcompeted by the autochthonous microflora in the soil. Here developments using specific carrier materials for the inoculum have been proven to be successful, which give inocula a protected initial niche for performance (van Elsas and Heijnen, 1990). Furthermore, an improved understanding of the ecology of inoculants is required for more reliable and efficient use (Berg et al., 2017).

Numerous experiments, in which treatments with *Trichoderma harzianum* (Wrede, 2015), cyanamid, stone dust and fertilizers amended with organic compounds such as humus, alkaline substances and seaweed were tested, were not or not sufficiently effective. Also, by incorporating spent mushroom compost, a composted substrate from mushroom production, into ARD soil, an increase of microbial activity was achieved leading to a significant increase in shoot growth, an effect being comparable to that of pasteurization (Manici, 2015; Franke-Whittle et al., 2018). More research is necessary to support or reject the hypothesis that many of these compounds were ineffective in harming soil pathogens, because they might have persisted in microhabitats different from those reached by the amendments.

The influence of important compounds exuded or released from roots ploughed into soil must be evaluated. In this respect, also carbon source-dependent effects of anaerobic soil disinfestation might be discussed (Hewavitharana and Mazzola, 2016).

Arbuscular mycorrhizal fungi (AMF) are essential endophytic players in the microbial network in the rhizosphere as well as in plant root systems. Besides the often-cited promotion of P acquisition, the mycorrhizal host plants have manifold advantages for their survival and productivity (Finlay, 2004; Smith and Read, 2008; Smith and Smith, 2012). Thus, a positive contribution of AMF to healthy apple growth and productivity is the normal case. The selection and application of AMF isolates for a recovery from ARD has been considered a possible strategy, but seems to be a rather difficult aim. Since long, this approach gained often less successful results in other instances of disease control, especially in field trials (Schönbeck et al., 1994; Linderman, 2000; Whipps, 2004). However, the AM symbiosis can lead to striking positive effects in plant productivity when damages of abiotic stress, e.g. drought (Pinior et al., 2005) or of infections by soil borne pathogens (Grunewaldt-Stöcker and von Alten, 2003; Whipps, 2004) and of nematodes (Calvet et al., 2001) were diminished. Regarding ARD, Čatská (1994) described a promising significant increase in productivity of apple plants (shoot and root biomass) due to *Glomus fasciculatum,* applied to ARD soil of two diverse soil types. Moreover, this mycorrhizal effect occurred together with an altered composition of the rhizosphere microbiome. Mehta and Bharat (2013) confirmed in tests with several AM fungi the specific success of a *Glomus fasciculatum* strain to overcome apple growth depression in ARD soil.

To apply selected effective AMF for an ARD therapy at a large scale in nurseries or field sites, the production of AMF inoculum, the formulation, shelf life and commercial supply are yet difficult (Azcón-Aguilar and Barea, 1997; Whipps, 2004). Nevertheless, the strategy to exploit naturally occurring or introduced AMF with a potential to alleviate abiotic stress and to control soil borne pathogens in combination with other biological agents or measurements against ARD seems attractive. AMF together with fine root endophytes (*Glomus tenue*, Orchard et al., 2017), are essential in the microbiome network and need attention, promotion and protection by all cultural practices. The recently widened molecular methods (e.g. real-time PCR quantification of AMF, Alkan et al., 2004, Voříšková et al., 2017; DNA based sequencing for identification and diversity studies of AMF, Vasar et al., 2017) can help to determine fungal communities with positive effects on apple plants. Also, AMF isolates harboring mycoviruses (Ikeda et al., 2012) or endobacteria (Venice et al., 2017) are of interest with regard to their influence on the symbiotic performance of AMF in ARD soil as well as in biocontrol strategies.

Besides modulating the soil microbiome, several recent strategies include the improvement of plant tolerance towards replant disease. Breeding of less susceptible rootstocks seems feasible as tolerant genotypes are available in *Malus* germplasm (e.g. Isutsa and Merwin, 2000; St. Laurent et al., 2010; Robinson et al., 2012; Volk et al., 2013). In addition, an improved strategy for defense responses of plant roots by modulating cellular signals such as the oscillation of Ca^{2+} concentration, reactive oxygen species burst or protein kinase activity (Emmett et al., 2014) is under debate.

Finally, more work needs to be done to assess the socio-economic benefits of such approaches.

Conclusions

Despite increasing data on ARD, combined efforts of plant scientists, ecologists, microbiologists, soil scientists as well as socio-economists and growers are needed to fully understand and overcome ARD. The German consortium BonaRes ORDIAmur (Overcoming Replant Disease by an Integrated Approach; www.ordiamur.de) aims at finding indicators for infected soil, to restore its functional biodiversity, to identify and use genetic factors controlling ARD in apple and to optimize the composition of microbial communities to promote apple growth in ARD soil (Figure 4). Two important considerations for future research in ARD have to be taken into account: Firstly, a proper comparison to healthy or virgin soil is difficult, since even small spatial distances to sites where healthy soil is taken might involve very drastic changes in soil physical, chemical and biological properties. Moreover, the vegetation of the control site will also influence the (micro)biome of the soils. On the other hand, disinfected soil can neither be considered a proper control soil. Secondly, future studies should emphasize the soil sampling and distinguish bulk soil, rhizosphere and rhizoplane, as well as define

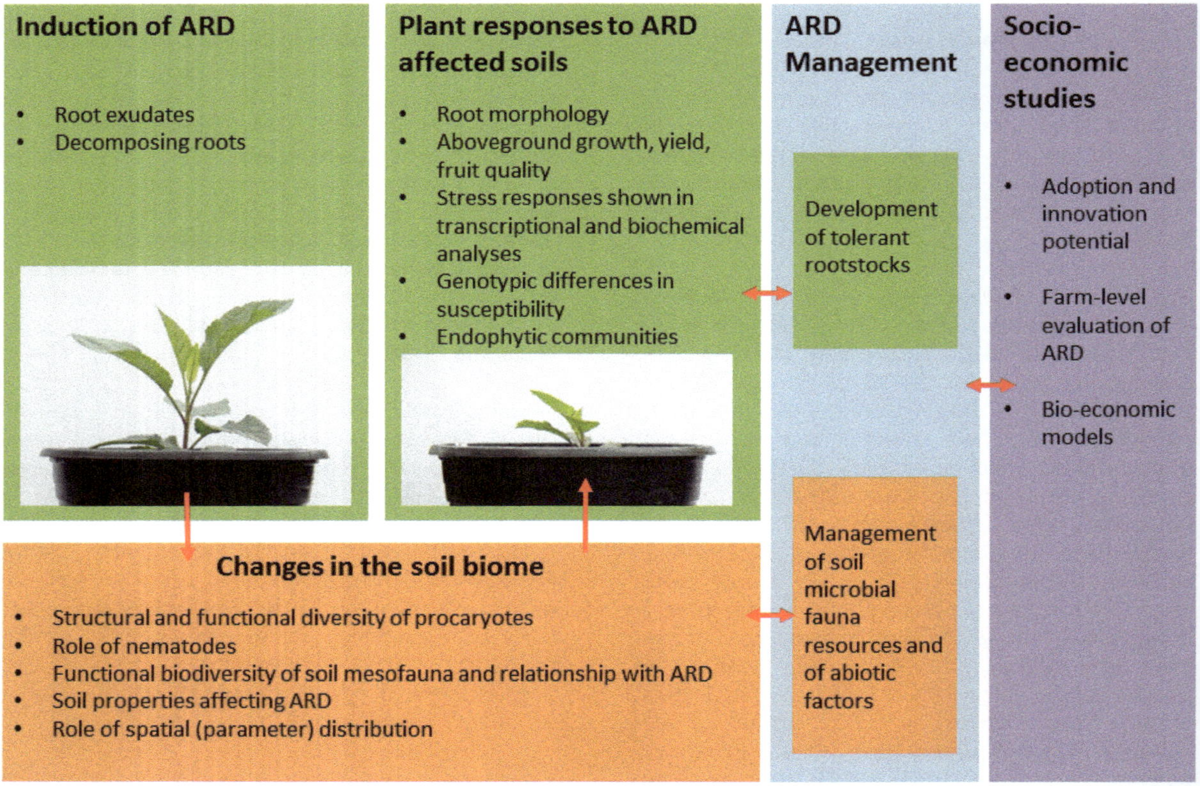

Figure 4. The understanding and management of apple replant disease (ARD) goes beyond disciplinary expertise - The ORDIAmur concept: Complex interactions of plant driven metabolites and soil-borne (micro)biome changes induce ARD. Future studies will have to focus on managing techniques including manipulation of plant attributes and soil microbial communities accompanied by socio-economic studies.

the root order and degree of damage. For both aspects, reproducible and internationally accepted definitions would be helpful.

Acknowledgements
The authors gratefully acknowledge funding of the project ORDIAmur by the German Federal Ministry of Research and Education within the frame of the program BonaRes (grant no. 031B0025).

References

Adam, M., Westphal, A., Hallmann, J., and Heuer, H. (2014). Invertebrate microbiology: Specific microbial attachment to root knot nematodes in suppressive soil. Environ. Microbiol. 80, 2679-2686. https://doi:10.1128/AEM.03905-13

Alkan, N., Gadkar, V., Coburn, J., Yarden, O., and Kapulnik, Y. (2004). Quantification of the arbuscular mycorrhizal fungus *Glomus intraradices* in host tissue using real-time polymerase chain reaction. New Phytol. 161, 877-885. doi:10.1111/j.1469-8137.2003.00975.x

Amelung, W., Brodowski, S., Sandhage-Hofmann, A., and Bol, R. (2008). Combining biomarker with stable isotope analyses for assessing the transformation and turnover of soil organic matter. Adv. Agron. 100, 155-250. https://doi.org/10.1016/S0065-2113(08)00606-8

Atucha, A., Emmett, B., and Bauerle, T.L. (2014). Growth rate of fine root systems influences rootstock tolerance to replant disease. Plant Soil 376, 337-346. https://doi.org/10.1007/s11104-013-1977-5

Ayyagari, J., Akkara, A., and Kaplan, D.L. (1996). Enzyme-mediated polymerization reactions: peroxidase-catalyzed polyphenol synthesis. Acta Polym. 47, 193-203. https://doi.org/10.1002/actp.1996.010470501

Azcón-Aguilar, C., and Barea, J. M. (1997). Applying mycorrhiza biotechnology to horticulture: significance and potentials. Sci. Hortic. 68, 1-24. https://doi.org/10.1016/S0304-4238(96)00954-5

Berg, G., Köberl, M, Rybakova, D., Müller, H., Grosch, R., and Smalla, K. (2017). Plant microbial diversity is suggested as the key to future biocontrol and health trends. FEMS Microbiology Ecology 93, fix050. https://doi:10.1093/femsec/fix050

Böllmann, J., Elmer, M., Woellecke, J., Raidl, S., and Huettl, R.F. (2010). Defensive strategies of soil fungi to prevent grazing by *Folsomia candida* (Collembola). Pedobiologia 53, 107-114. https://doi.org/10.1016/j.pedobi.2009.06.003

Bogena, H.R., Herbst, M., Huisman, J.A., Rosenbaum, U., Weuthen, A., and Vereecken, H. (2010). Potential of wireless sensor networks for measuring soil water content variability. Vadose Zone J. 9, 1002-1013. https://doi.org/10.2136/vzj2009.0173

Börner, H. (1959). The apple replant problem. I. The excretion of phloridzin from apple root residues. Contributions of the Boyce Thompson Institute of Plant Research 20, 39-56.

Brown, P.D., Morra, M.J., McCafferey, J.P., Auld, D.L., and Williams, L. (1991). Allelochemicals produced during glucosinolate degradation in soil. J. Chem. Ecol. 17, 2021-2034. https://doi.org/10.1007/BF00992585

Calvet, C., Pinochet, J., Hernández-Dorrego, A., Estáun, V., and Camprubi, A. (2001). Field microplot performance of the peach-almond hybrid GF-677 after inoculation with arbuscular mycorrhizal fungi in a replant soil infested with root-knot nematodes. Mycorrhiza 10, 295-300. https://doi.org/10.1007/PL00009998

Caputo, F., Nicoletti, F., De Luca Picione, F., and Manici, L.M. (2015). Rhizospheric changes of fungal and bacterial communities in relation to soil health of multi-generation apple orchards. Biol. Control 88, 8-17. https://doi.org/10.1016/j.biocontrol.2015.04.019

Carpenter, R.J., Hartzell, J.D., Forsberg, J.A., Babel, B.S., and Ganesan, A. (2008). *Pseudomonas putida* war wound infection in a US Marine: a case report and review of the literature. J. Infection 56(4), 234-240. https://doi.org/10.1016/j.jinf.2008.01.004

Caruso, F.L., Neubauer, B.F., and Begin, M.D. (1989). A histological study of apple roots affected by replant disease. Can. J. Bot. 67, 742-749. https://doi.org/10.1139/b89-100

Čatská, V. (1994). Interrelationships between vesicular-arbuscular mycorrhiza and rhizosphere microflora in apple replant disease. Biol. Plantarum 36, 99-104. https://doi.org/10.1007/BF02921276

Chelinho, S., Domene, X., Andrés, P., Natal-da-Luz, T., Norte, C., Rufino, C., Lopes, I., Cachada, A., Espindola, E., Ribeiro, R., Costa-Duarte, A., and Sousa, J.P. (2014). Soil microarthropod community testing: A new approach to increase the ecological relevance of effect data for pesticide risk assessment. Applied Soil Ecol. 83, 200-209. https://doi.org/10.1016/j.apsoil.2013.06.009

Ciglasch, H., Busche, J., Amelung, W., Totrakool, S., and Kaupenjohann, M. (2008). Field aging of insecticides after repeated application to a northern Thailand ultisol. J. Agr. Food Chem. 56, 9555-9562. https://doi.org/10.1021/jf801545h

Covey Jr, R.P., Benson, N.R., and Haglund, W.A. (1979). Effect of soil fumigation on the apple replant disease in Washington [USA]. Phytopathology 69, 684-686.

Derrien, D., Marol, C., Balabane, M., and Balesdent, J. (2006). The turnover of carbohydrate content in a cultivated soil estimated by 13C natural abundances. Eur. J. Soil Sci. 57(4), 547-557. https://doi.org/10.1111/j.1365-2389.2006.00811.x

Djigal, D., Brauman, A., Diop, T.A., Chotte, J.L., and Villenave, C. (2004). Influence of bacterial-feeding nematodes (Cephalobidae) on soil microbial communities during maize growth. Soil Biol. Biochem. 36, 323-331. https://doi.org/10.1016/j.soilbio.2003.10.007

Dunn, R.A., and Mai, W.F. (1972). Root diseases of fruit trees in New York State. V. Growth of apple trees in response to preplant treatment with nematicides and a broad-spectrum fumigant. Plant Dis. Rep. 56, 577-580.

Elhady, A., Giné, A., Topalovic, O., Jacquiod, S., Sørensen, S.J., Sorribas, F.J., and Heuer, H. (2017). Microbiomes associated with infective stages of root-knot and lesion nematodes in soil. Plos One 12, e0177145. https://doi.org/10.1371/journal.pone.0177145

Emmett, B., Nelson, E.B., Kessler, A., and Bauerle, T.L. (2014). Fine-root system development and susceptibility to pathogen colonization. Planta 239, 325-340. https://doi.org/10.1007/s00425-013-1989-7

Fan, H., Zhao, Z., Liu, H., Guodong, Z.X., and Zhang, Z. (2010). Changes of soil nutrition in root zone and their effects on growth of the replanted apple. Yuanyi Xuebao 35, 1727-1734 (in Chinese; cited from abstract).

Fazio, G., Kviklys, D., Grusak, M.A., and Robinson, T. (2012). Soil pH, soil type and replant disease affect growth and nutrient absorption in apple rootstocks. New York State Fruit Quarterly 20, 22-28.

Finlay, R.D. (2004). Mycorrhizal fungi and their multifunctional roles. Mycologist 18, 91-96. https://doi.org/10.1017/S0269-915X(04)00205-8

Frampton, G.K. (2002). Long-term impacts of an organophosphate-based regime of pesticides on field and field-edge Collembola communities. Pest Manag. Sci. 58, 991-1001. https://doi.org/10.1002/ps.580

Franke-Whittle, I.H., Manici, L.M., Insam, H., and Stres, B. (2015). Rhizosphere bacteria and fungi associated with plant growth in soils of three replanted apple orchards. Plant Soil 395, 317-333. https://doi.org/10.1007/s11104-015-2562-x

Franke-Whittle, I.H., Fernández-Delgado Juárez, M., Insam, H., Schweizer, S., Naef, A., Topp, A.R., Kelderer, M., Rühmer, T., Baab, G., Henfrey, J., and Manici, L.M. (2018): Performance evaluation of locally available composts to reduce replant disease in apple orchards of central Europe. Renewable Agriculture and Food Systems. https://doi.org/10.1017/S1742170518000091

Freckman, D.W. (1988). Bacterivorous nematodes and organic-matter decomposition. Agr. Ecosyst. Environ. 24, 195-217. https://doi.org/10.1016/0167-8809(88)90066-7

Gebbers, R., and Adamchuk, V.I. (2010). Precision agriculture and food security. Science 327, 828-831. https://doi.org/10.1126/science.1183899

Gebremikael, M.T., Steel, H., Buchan, D., Bert, W., and de Neve, S. (2016). Nematodes enhance plant growth and nutrient uptake under C and N-rich conditions. Sci. Rep.-UK 6, 32862. https://doi.org/10.1038/srep32862

Glisczynski, F., Sandhage-Hofmann, A., Amelung, W., and Pude, R. (2016). Biochar-compost substrates do not promote growth and fruit quality of a replanted German apple orchard with fertile Haplic Luvisol soils. Sci. Hortic. 213, 110-114. https://doi.org/10.1016/j.scienta.2016.10.023

Grunewaldt-Stöcker, G., and von Alten, H. (2003). Plant health effects of Acremonium root endophytes compared to those of arbuscular mycorrhiza. In Roots: The dynamic interface between plants and the earth, J. Abe, ed. (Dordrecht, Netherlands: Kluwer Academic Publishers), pp. 445-454. https://doi.org/10.1007/978-94-017-2923-9_44

Hai-Feng, X., Gen, L.I., Da-Ming, L.I., Feng, H.U., and Hui-Xin, L.I. (2014). Effect of different bacterial-feeding nematode species on soil bacterial numbers, activity, and community composition. Pedosphere 24, 116-124. https://doi.org/10.1016/S1002-0160(13)60086-7

Hatzinger, P.B., and Alexander, M. (1995). Effect of aging of chemicals in soils on their biodegradability and extractability. Environ. Sci. Technol. 29, 537-545. https://doi.org/10.1021/es00002a033

Henfrey, J.L., Baab, G., and Schmitz, M. (2015). Physiological stress responses in apple under replant conditions. Sci. Hortic.-Amsterdam 194, 111-117. https://doi.org/10.1016/j.scienta.2015.07.034

Herbst, M., Bornemann, L., Graf, A., Welp, G., Vereecken, H., and Amelung, W. (2012). A geostatistical approach to the field-scale pattern of heterotrophic soil CO_2 emission using covariates. Biogeochem. 111, 377-392. https://doi.org/10.1007/s10533-011-9661-4

Hewavitharana, SS., and Mazzola, M. (2016). Carbon Source-Dependent Effects of Anaerobic Soil Disinfestation on Soil Microbiome and Suppression of *Rhizoctonia solani* AG-5 and *Pratylenchus penetrans*. Phytopathology 106, 1015-1028. https://doi.org/10.1094/PHYTO-12-15-0329-R

Hoestra, H. (1968). Replant diseases of apple in The Netherlands. Dissertation Landbouwhogesch, Wageningen. https://edepot.wur.nl/285366

Hoestra, H., and Oostenbrink, M. (1962). Nematodes in relation to plant growth. IV. *Pratylenchus penetrans* (Cobb) on orchard trees. Neth. J. Agr. Sci. 10, 286-296.

Hofmann, A., Wittenmayer, L., Arnold, G., Schieber, A., and Merbach, W. (2009). Root exudation of phloridzin by apple seedlings (*Malus x domestica* Borkh.) with symptoms of apple replant disease. J. Appl. Bot. Food Qual. 82, 193-198.

Hopkin, S.P. (1997). Biology of the springtails. (Insecta: Collembola). (Oxford, New York, USA: Oxford University Press). http://www.loc.gov/catdir/enhancements/fy0605/96034592-d.html

Ikeda, Y., Shimura, H., Kitahara, R., Masuta, C., and Ezawa, T. (2012). A novel virus-like double-stranded RNA in an obligate biotroph arbuscular mycorrhizal fungus: A hidden player in mycorrhizal symbiosis. MPMI 25, 1005-1012. https://doi.org/10.1094/MPMI-11-11-0288

Innocenti, G., Ganassi, S., Montanari, M., Branzanti, M.B., and Sabatini, M.A. (2009). Response of plant growth to Collembola, arbuscular mycorrhizal and plant pathogenic fungi interactions. B. Insectol. 62, 191-195.

Innocenti, G., Montanari, M., Ganassi, S., and Sabatini, M.A. (2011). Does substrate water content influence the effect of Collembola-pathogenic fungus interaction on plant health? A mesocosm study. B. Insectol. 64, 73-76.

Isutsa, D.K., and Merwin, I.A. (2000). Malus germplasm varies in resistance or tolerance to apple replant disease in a mixture of New York orchard soils. Hortscience 35(2), 262-268.

Jaffee, B.A., Abawi, G.S., and Mai, W.F. (1982). Role of soil microflora and *Pratylenchus penetrans* in an apple replant disease. Phytopathology 72, 247-251. https://doi.org/10.1094/Phyto-72-247

Jonkers, H., Hoestra, H., Borsboom, O., and Power, A. (1980). Soil pH in fruit trees in relation to specific apple replant disorder (SARD). II: The first five years at the Wageningen research plot. Sci. Hortic.-Amsterdam 13, 149-154. https://doi.org/10.1016/0304-4238(80)90079-5

Kanters, C., Anderson, I.C., and Johnson, D. (2015). Chewing up the wood-wide web: Selective grazing on ectomycorrhizal fungi by Collembola. Forests 6, 2560-2570. http://doi.org/10.3390/f6082560

Klaus, H. (1939). Das Problem der Bodenmüdigkeit unter Berücksichtigung des Obstbaus. Landw. Jahrb. 89, 413-459.

Kögel-Knabner, I., and Amelung, W. (2014). Dynamics, chemistry, and preservation of organic matter in soils. In Treatise on Geochemistry (Second Edition), H.D. Holland and K.K. Turekian, eds. (Oxford, UK: Elsevier Ltd), pp. 157-215.

Kuzyakov, Y., and Larionova, A.A. (2005). Root and rhizomicrobial respiration: A review of approaches to estimate respiration by autotrophic and heterotrophic organisms in soil. J. Plant Nutr. Soil Sci. 168, 503-520. http://doi.org/10.1002/jpln.200421703

Larsen, T., Schjonning, P., and Axelsen, J. (2004). The impact of soil compaction on euedaphic Collembola. Appl. Soil Ecol. 26, 273-281. https://doi.org/10.1016/j.apsoil.2003.12.006

Lartey, R.T., Curl, E.A., and Peterson, C.M. (1994). Interactions of mycophygous Collembola and biological control fungi in the suppression of *Rhizoctonia solani*. Soil Biol. Biochem. 26, 81-88. https://doi.org/10.1016/0038-0717(94)90198-8

Leisso, R., Rudell, D., and Mazzola, M. (2017). Metabolic composition of apple rootstock rhizodeposits differs in a genotype-specific manner and affects growth of subsequent plantings. Soil Biol. Biochem. 113, 201-214. https://doi.org/10.1016/j.soilbio.2017.06.011

Linderman, R.G. (2000). Effects of mycorrhizas on plant tolerance to diseases. In Arbuscular mycorrhizas: Physiology and Function, Y. Kapulnik and D.D. Douds Jr, eds. (Dordrecht, Netherlands: Kluwer Academic Publishers), pp. 345-365. https://doi.org/10.1007/978-94-017-0776-3_15

Liu, E.T., Wang, G.S., Li, Y.Y., Shen, X., Chen, X.S., Song, F.H., Wu, S.J., Chen, Q., and Mao, Z.Q. (2014). Replanting affects the tree growth and fruit quality of Gala apple. J. Integr. Agr. 13, 1699-1706. https://doi.org/10.1016/S2095-3119(13)60620-6

Lueking, A.D., Huang, W., Soderstrom-Schwarz, S., and Kim Jr, M.S. (2000). Relationship of soil organic matter characteristics to organic contaminant sequestration and bioavailability. J. Environ. Qual. 29, 317-323. https://doi.org/10.2134/jeq2000.00472425002900010040x

Mai, W.F., and Abawi, G.S. (1978). Determining the cause and extent of apple, cherry, and pear replant diseases under controlled conditions. Phytopathology 68, 1540-1544. https://doi.org/10.1094/Phyto-68-1540

Mai, W.F., and Abawi, G.S. (1981). Controlling replant disease of stone fruits in northeastern United States by preplant fumigation. Plant Dis. 65, 859-864. https://doi.org/10.1094/PD-65-859

Mai, W.F., Merwin, I.A., and Abawi, G.S. (1994). Diagnosis, etiology and management of replant disorders in New York cherry and apple orchards. Acta Hortic. 363, 33-42. https://doi.org/10.17660/ActaHortic.1994.363.5

Manici, L (2015). Bio-Incrop CORE organic - Innovative cropping technique to increase soil health in organic fruit tree crops. EU project 2012 - 2014. Project Leader: Manici, L.M. http://orgprints.org/29112/1/BIO-INCROP_COII_final_report_web_version.pdf

Manici, L.M., Caputo, F., and Saccà, M.L. (2017). Secondary metabolites released into the rhizosphere by *Fusarium oxysporum* and *Fusarium* spp. as underestimated component of nonspecific replant disease. Plant Soil. https://doi.org/10.1007/s11104-016-3152-2

Manici, L.M., Kelderer, M., Franke-Whittle, I.H., Rühmer, T., Baab, G., Nicoletti, F., Caputo, F., Topp, A., Insam, H., and Naef A. (2013). Relationship between root-endophytic microbial communities and replant disease in specialized apple growing areas in Europe. Appl. Soil Ecol. 72, 207-214. https://doi.org/10.1016/j.apsoil.2013.07.011

Manici, L. M., Kelderer, M., Caputo, F., Saccà, M. L., Nicoletti, F., Topp, A. R., and Mazzola, M. (2018). Involvement of *Dactylonectria* and *Ilyonectria* spp. in tree decline affecting multi-generation apple orchards. Plant and Soil 425: 217-230. https://doi.org/10.1007/s11104-018-3571-3

Mazzola, M. (1997). Identification and pathogenicity of *Rhizoctonia* spp. isolated from apple roots and orchard soils. Phytopathology 87(6), 582-587. https://doi.org/10.1094/PHYTO.1997.87.6.582

Mazzola, M. (1998). Elucidation of the microbial complex having a causal role in the development of apple replant disease in Washington. Phytopathology 88, 930-938. https://doi.org/10.1094/PHYTO.1998.88.9.930

Mazzola, M. (2007). Manipulation of rhizosphere bacterial communities to induce suppressive soils. J. Nematol. 39(3), 213-220.

Mazzola, M., Brown, J., Izzo, A.D., and Cohen, M.F. (2007). Mechanism of action and efficacy of seed meal-induced pathogen suppression differ in a Brassicaceae species and time dependent manner. Phytopathology 97, 454-460. https://doi.org/10.1094/PHYTO-97-4-0454

Mazzola, M., Granatstein, D.M., Elfving, D.C., and Mullinix, K. (2001). Suppression of specific apple root pathogens by *Brassica napus* seed meal amendment regardless of glucosinolate content. Phytopathology 91, 673-79. https://doi.org/10.1094/PHYTO.2001.91.7.673

Mazzola, M., Granatstein, D.M., Elfving, D.C., Mullinix, K., and Gu, Y.H. (2002) Cultural management of microbial community structure to enhance growth of apple in replant soils. Phytopathology 92(12), 1363-1366. https://doi.org/10.1094/PHYTO.2002.92.12.1363

Mazzola, M., and Gu, Y.H. (2000). Impact of wheat cultivation on microbial communities from replant soils and apple growth in greenhouse trials. Phytopathology 90(2), 114-119. https://doi.org/10.1094/PHYTO.2000.90.2.114

Mazzola, M., Hewavitharana, S.S., and Strauss, S.L. (2015). Brassica seed meal soil amendments transform the rhizosphere microbiome and improve apple production through resistance to pathogen reinfestation. Phytopathology 105, 460-469. https://doi.org/10.1094/PHYTO-09-14-0247-R

Mazzola, M., and Manici, L.M. (2012). Apple replant disease: Role of microbial ecology in cause and control. Annu. Rev. Phytopathol. 50, 45-65. https://doi.org/10.1146/annurev-phyto-081211-173005

Mazzola, M., and Mullinix, K. (2005). Comparative field efficacy of management strategies containing *Brassica napus* seed meal or green manure for the control of apple replant disease. Plant Dis. 89, 1207-1213. https://doi.org/10.1094/PD-89-1207

Mehta, P., and Bharat, N.K. (2013). Effect of indigenous arbuscular- mycorrhiza (*Glomus* spp) on apple (*Malus domestica*) seedlings grown in replant diseased soil. Ind. J. Agr. Sci. 83, 1173-1178.

Nicola, L., Turco, E., Albanese, D., Donati, C., Thalheimer, M., Pindo, M., Insam, H., Cavalieri, D., and Pertot, I. (2017). Fumigation with dazomet modifies soil microbiota in apple orchards affected by replant disease. Appl. Soil Ecol. 113, 71-79. https://doi.org/10.1016/j.apsoil.2017.02.002

Nitt, H., Yim, B., Wrede, A., Hanschen, F., Loesing, H., Schreiner, M., Smalla, K., Grunewaldt, J., and Winkelmann, T. (2015). Bodenmüdigkeit und Lösungen (Baumschulen) Gesunder Boden - Gesunder Gartenbau. Kongress zum Jahr des Bodens, Tagungsband, 16. November 2015, Berlin. https://www.bmel.de/SharedDocs/Downloads/Landwirtschaft/JahrdesBodens/TagungsbandKongressJahrdesBodens.pdf?__blob=publicationFile

Oades, J.M. (1988). The retention of organic matter in soils. Biogeochem. 5, 35-70. https://doi.org/10.1007/BF02180317

Orchard, S., Standish, R.J., Dickie, I.A., Renton, M., Walker, C., Moot, D. and Ryan, M.H. (2017). Fine root endophytes under scrutiny: a review of the literature on arbuscule-producing fungi recently suggested to belong to the Mucoromycotina. Mycorrhiza 27, 619-638. https://doi.org/10.1007/s00572-017-0782-z

Otto, G., Winkler, H., and Szabo K. (1994). Proof of actinomycetes in rootlets of species of *Rosaceae* from a SARD soil - a contribution to the specificity of replant diseases. Acta Hortic. 363, 43-48. https://doi.org/10.17660/ActaHortic.1994.363.6

Peruzzi, E., Franke-Whittle, I.H., Kelderer, M., Ciavatta, C., and Insam, H. (2017). Microbial indication of soil health in apple orchards affected by replant disease. Appl. Soil Ecol. 119, 115-127. https://doi.org/10.1016/j.apsoil.2017.06.003

Pinior, A., Grunewaldt-Stöcker, G., von Alten, H., and Strasser, R.J. (2005). Mycorrhizal impact on drought stress tolerance of rose plants probed by chlorophyll a fluorescence, proline content and visual scoring. Mycorrhiza 15, 596-605. https://doi.org/10.1007/s00572-005-0001-1

Robinson, T., Autio, W., Clements, J., Cowgill, W., Embree, C., Gonzalez, V., Hoying, S., Kushad, M., Parker, M., Parra, R., and Schupp, J. (2012). Rootstock tolerance to apple replant disease for improved sustainability of apple production. Acta Hortic. 940, 521-528.

Rosendahl, I., Siemens, J., Kindler, R., Groeneweg, J., Zimmermann, J., Czerwinski, S., Lamshöft, M., Laabs, V., Wilke, B.M., Vereecken, H., and Amelung, W. (2012). Persistence of the fluoroquinolone antibiotic difloxacin in soil and lacking effects on N-turnover. J. Environ. Qual. 41, 1275-1283. https://doi.org/10.2134/jeq2011.0459

Roy, S., Bano, R., Saxena, P., and Bhatt, R.K. (2014). Land uses and its impact on community structure of soil collembola. Range Manag. Agrofor. 35, 27-31.

Rumberger, A., Merwin, I.A., and Thies, J.E. (2007). Microbial community development in the rhizosphere of apple trees at a replant disease site. Soil Biol. Biochem. 39, 1645-1654. https://doi.org/10.1016/j.soilbio.2007.01.023

Sabatini, M.A., and Innocenti, G. (2000). Functional relationships between Collembola and plant pathogenic fungi of agricultural soils. Pedobiologia 44, 467-475. https://doi.org/10.1078/S0031-4056(04)70064-5

Sabatini, M.A., and Innocenti, G. (2001). Effects of Collembola on plant-pathogenic fungus interactions in simple experimental systems. Biol. Fert. Soils 33, 62-66. https://doi.org/10.1007/s003740000290

San Miguel, A., Raveton, M., Lemperiere, G., and Ravanel, P. (2008). Phenylpyrazoles impact on *Folsomia candida* (Collembola). Soil Biol. Biochem. 40, 2351-2357. https://doi.org/10.1016/j.soilbio.2008.05.014

Savory, B.M. (1966). Specific replant diseases causing root necrosis and growth depression in perennial fruit and plantation crops. Research Review No. 1. Commonwealth Bureau of Horticulture and Plantation Crops, East Malling, Maidstone, Kent, England.

Schönbeck, F., Grunewaldt-Stöcker, G., and von Alten, H. (1994). Mycorrhizae. In Epidemiology and management of root diseases, C.L. Campbell and D.M. Benson, eds. (Berlin, Heidelberg, New York: Springer Verlag), pp. 65-81.

Schrader, S., Langmaack, M., and Helming, K. (1997). Impact of Collembola and enchytraeidae on soil surface roughness and properties. Biol. Fert. Soils 25, 396-400. https://doi.org/10.1007/s003740050331

Sewell, G.W., Preece, D.A., and Elsey, R.F. (1988). Apple replant disease: the influence of soil phosphorus and other factors on the growth responses of apple seedlings to soil fumigation with chloropicrin. Ann. Appl. Biol. 113, 605-615. https://doi.org/10.1111/j.1744-7348.1988.tb03338.x

Shin, S., Lv, J., Fazio, G., Mazzola, M., and Zhu, Y. (2014). Transcriptional regulation of ethylene and jasmonate mediated defense response in apple (*Malus domestica*) root during *Pythium ultimum* infection. Horticulture Research 1, 14053. https://doi.org/10.1038/hortres.2014.53

Shin, S., Zheng, P., Fazio, G., Mazzola, M., Main, D., and Zhu, Y. (2016). Transcriptome changes specifically associated with apple (*Malus domestica*) root defense response during *Pythium ultimum* infection. Physiol. Mol. Plant P. 94, 16-26. https://doi.org/10.1016/j.pmpp.2016.03.003

Smith, S.E., and Read, D.J. (2008). Mycorrhizal symbiosis (Third Edition, New York, USA: Academic Press).

Smith, S.E., and Smith, F. (2012). Fresh perspectives on the roles of arbuscular mycorrhizal fungi in plant nutrition and growth. Mycologia 104, 1-13. https://doi.org/10.3852/11-229

Spath, M., Insam, H., Peintner, U., Kelderer, M., Kuhnert-Finkernagel, R., and Franke-Whittle I.H. (2015). Linking soil biotic and abiotic factors to apple replant disease: a greenhouse approach. J. Phytopathol. 163, 287-299. https://doi.org/10.1111/jph.12318

Spethmann, W., and Otto, G. (2003). Replant problems and soil sickness. In Encyclopedia of Rose Science, A.V. Roberts, T. Debener, S. Gudin, eds. (Oxford, UK: Elsevier LTD), pp. 169-180.

Steinaker, D.F., and Wilson, S.D. (2008). Scale and density dependent relationships among roots, mycorrhizal fungi and collembola in grassland and forest. OIKOS 117, 703-710. https://doi.org/10.1111/j.2008.0030-1299.16452.x

St. Laurent, A., Merwin, I.A., Fazio, G., Thies, J.E., and Brown, M.G. (2010). Rootstock genotype succession influences apple replant disease and root-zone microbial community composition in an orchard soil. Plant Soil 337, 259-72. https://doi.org/10.1007/s11104-010-0522-z

St. Laurent, A., Merwin, I.A., and Thies, J.E. (2008). Long-term orchard groundcover management affect soil microbial communities and apple replant disease severity. Plant Soil 304, 209-225. https://doi.org/10.1007/s11104-008-9541-4

Sun, J., Zhang, Q., Zhou, J., and Wei, Q. (2014). Illumina amplicon sequencing of 16S rRNA tag reveals bacterial community development in the rhizosphere of apple nurseries at a replant disease site and a new planting site. Plos One 9, e111744. https://doi.org/10.1371/journal.pone.0111744

Tewoldemedhin, Y.T., Mazzola, M., Labuschagne, I., and McLeod, A. (2011). A multi-phasic approach reveals that apple replant disease is caused by multiple biological agents, with some agents acting synergistically. Soil Biol. Biochem. 43, 1917-1927. https://doi.org/10.1016/j.soilbio.2011.05.014

Tukey, R.B., Dow, A.I., and Halvorson, A.R. (1984). Fertilizer guide - Fruit trees for entire state. Publication FG 0028a. College of Agriculture Washington State University, Pullman, Washington.

Utkhede, R.S. (2006). Soil sickness, replant problem or replant disease and its integrated control. Allelopathy J. 18, 23-38.

Utkhede, R.S., and Smith, E.M. (1992). Promotion of apple tree growth and fruit production by the EBW-4 strain of *Bacillus subtilis* in apple replant disease soil. Can. J. Microbiol. 38(12), 1270-1273. https://doi.org/10.1139/m92-209

Utkhede, R.S., Vrain, T.C., and Yorston J.M. (1992). Effects of nematodes, fungi, and bacteria on the growth of young apple trees grown in apple replant disease soils. Plant Soil 139, 1-6. http://www.jstor.org/stable/42937909

Van Elsas, J.D., and Heijnen, C.E. (1990). Methods for the introduction of bacteria into soil: a review. Biol. Fert. Soils 90, 127-133. https://doi.org/10.1007/BF00336248

Van Schoor, L., Denman, S., and Cook, N.C. (2009). Characterisation of apple replant disease under South African conditions and potential biological management strategies. Sci. Hortic.-Amsterdam 119, 153-162. https://doi.org/10.1016/j.scienta.2008.07.032

Vasar, M., Andreson, R., Davison, J., Jairus, T., Moora, M., Remm, M., Young, J.P.W., Zobel, M., and Öpik, M. (2017). Increased sequencing depth does not increase captured diversity of arbuscular mycorrhizal fungi. Mycorrhiza 27, 761-773. https://doi.org/10.1007/s00572-017-0791-y

Venice, F., de Pinto, M.C., Novero, M., Ghignone, S., Salvioli, A., and Bonfante, P. (2017). *Gigaspora margarita* with and without its endobacterium shows adaptive responses to oxidative stress. Mycorrhiza 27, 747-759. https://doi.org/10.1007/s00572-017-0790-z

Volk, G.M., Chao, C.T., Norelli, J., Brown, S.K., Fazio, G., Peace, C., McFerson, J., Zhong, G.Y., and Bretting, P. (2015). The vulnerability of US apple (Malus) genetic resources. Genet. Resour. Crop Ev. 62, 765-794. https://doi.org/10.1007/s10722-014-0194-2

Volk, G.M., Henk, A.D., Richards, C.M., Forsline, P.L., and Chao, C.T. (2013): *Malus sieversii*. A Diverse Central Asian Apple Species in the USDA-ARS National Plant Germplasm System. HortScience 48, 1440-1444.

Von Bronsart, H. (1949). Der heutige Stand unseres Wissens von der Bodenmüdigkeit. J. Plant Nutr. Soil Sc. 45, 166-193. https://doi.org/10.1002/jpln.19490450116

Von Lützow, M., Kögel-Knabner, I., Ekschmitt, K., Matzner, E., Guggenberger, G., Marschner, B., and Flessa, H. (2006). Stabilization of organic matter in temperate soils: Mechanisms and their relevance under different soil conditions - A review. Eur. J. Soil Sci. 57, 426-445. https://doi.org/10.1111/j.1365-2389.2006.00809.x

Voříšková, A., Jansa, J., Püschel, D., Krüger, M., Cajthaml, T., Vosátka, M., and Janoušková, M. (2017) Real-time PCR quantification of arbuscular mycorrhizal fungi: does the use of nuclear or mitochondrial markers make a difference? Mycorrhiza 27, 577-585. https://doi.org/10.1007/s00572-017-0777-9

Vukicevich, E., Lowery, T., Bowen, P., Urbez-Torres, J. R., and Hart, M. (2016). Cover crops to increase soil microbial diversity and mitigate decline in perennial agriculture. A review. Agronomy for Sustainable Development 36 (3). https://doi.org/10.1007/s13593-016-0385-7

Walter, D.E., and Proctor, H.C. (2013). Mites. Ecology, Evolution and Behaviour; Life at a Microscale (2nd edition, Dordrecht, Netherlands: Springer).

Wang, K.H., and McSorley, R. (2005). Effects of soil ecosystem management on nematode pests, nutrient cycling, and plant health. APSnet Features, 2005-0105. https://doi.org/10.1094/APSnetFeatures/2005-0105

Wardle, D.A. (2006). The influence of biotic interactions on soil biodiversity. Ecol. Lett. 9, 870-886. https://doi.org/10.1111/j.1461-0248.2006.00931.x

Wardle, D.A. (2013). Communities and ecosystems. Linking the aboveground and belowground components. Princeton: Princeton University Press (Monographs in Population Biology). http://gbv.eblib.com/patron/FullRecord.aspx?p=1113398

Weiß, S., Bartsch, M., and Winkelmann, T. (2017a). Transcriptomic analysis of molecular responses in *Malus domestica* 'M26' roots affected by apple replant disease. Plant Mol. Biol. 94, 303-318. https://doi.org/10.1007/s11103-017-0608-6

Weiß, S., Liu, B., Reckwell, D., Beerhues, L., and Winkelmann, T. (2017b). Impaired defense reactions in apple replant disease-affected roots of *Malus domestica* 'M26'. Tree Physiol. https://doi.org/10.1093/treephys/tpx108

Whipps, J.M. (2004). Prospects and limitations for mycorrhizas in biocontrol of root pathogens. Can. J. Bot. 82, 1198-1227. https://doi.org/10.1139/b04-082

Wiesenberg, G.L.B., Schwarzbauer, J., Schmidt, M.W.I., and Schwark, L. (2004). Sources and turnover of organic matter in agricultural soils derived from n-alkane/n-carboxylic acid compositions and C-isotope signatures. Org. Geochem. 35, 1371-1393. https://doi.org/10.1016/j.orggeochem.2004.03.009

Willett, M., Smith, T.J., Peterson, A.B., Hinman, H., Stevens, R.G., Ley, T., Tvergyak, P., Williams, K.M., Maib, K.M., and Williams, J.W. (1994). Growing Profitable Apple Orchards in Replant Sites: An Interdisciplinary Team Approach in Washington State. HortTechnology 4, 175-181.

Wittenmayer, L., and Szabó, K. (2000). The role of root exudates in specific apple (*Malus* x *domestica* Borkh.) replant disease (SARD). J. Plant Nutr. Soil Sc. 163, 399-404. https://doi.org/10.1002/1522-2624(200008)163:4<399::AID-JPLN399>3.0.CO;2-8

Wrede, A. (2015). Bodenmüdigkeit bei Rosen - Was bringt Zugabe von *Trichoderma harzianum*? Deutsche Baumschule 11, 37-39.

Yim, B., Hanschen, F.S., Wrede, A., Nitt, H., Schreiner, M., Smalla, K., and Winkelmann, T. (2016). Effects of biofumigation using *Brassica juncea* and *Raphanus sativus* in comparison to disinfection using Basamid on apple plant growth and soil microbial communities at three field sites with replant disease. Plant Soil 406, 389-408. https://doi.org/10.1007/s11104-016-2876-3

Yim, B., Nitt, H., Wrede, A., Jacquiod, S., Sørensen, J., Winkelmann, T., and Smalla, K. (2017). Effects of soil pre-treatment with Basamid® granules, *Brassica juncea, Raphanus sativus* and *Tagetes patula* on bacterial and fungal communities at two replant disease sites. Front. Microbiol. 8, 1604. https://doi.org/10.3389/fmicb.2017.01604

Yim, B., Smalla, K., and Winkelmann, T. (2013). Evaluation of apple replant problems based on different soil disinfection treatments-links to soil microbial community structure? Plant Soil 366, 617-631. https://doi.org/10.1007/s11104-012-1454-6

Yim, B., Winkelmann, T., Ding, G.C., and Smalla, K. (2015). Different bacterial communities in heat and gamma irradiation treated replant disease soils revealed by 16S rRNA gene analysis - contribution to improved aboveground apple plant growth? Front. Microbiol. 6, 1224. https://doi.org/10.3389/fmicb.2015.01224

Yin, C., Xiang, L., Wang, G., Wang, Y., Shen, X., Chen, X., and Mao, Z. (2016). How to plant apple trees to reduce replant disease in apple orchard: A study on the phenolic acid of the replanted apple orchard. Plos One 12, e0167347. https://doi.org/10.1371/journal.pone.0167347

Yin, C., Xiang, L., Wang, G., Wang, Y., Shen, X., Chen, X., and Mao, Z. (2017). Phloridzin promotes the growth of *Fusarium moniliforme* (*Fusarium verticillioides*). Sci. Hortic.-Amsterdam 214, 187-194. https://doi.org/10.1016/j.scienta.2016.11.035

Zhu, Y., Fazio, G., and Mazzola, M. (2014). Elucidating the molecular responses of apple rootstock resistant to ARD pathogens: challenges and opportunities for development of genomics-assisted breeding tools. Horticulture Research 1, 14043. https://doi.org/10.1038/hortres.2014.43